Donald Frederick

and

Mildred Topp Othmer

Donald Frederick Othmer and Mildred Topp Othmer, wedding photograph, November 1950.

DONALD FREDERICK

AND

MILDRED TOPP OTHMER

A Commemorative of Their Lives and Legacies

EDITORS

Arnold Thackray
Amy Beth Crow

CHEMICAL HERITAGE FOUNDATION

PHILADELPHIA

Printed in the United States of America.

Design and layout by Patricia Wieland
Printed by United Book Press, Inc.

For information about CHF publications write
Chemical Heritage Foundation
315 Chestnut Street
Philadelphia, PA 19106-2702, USA
Fax: (215) 925-1954

Photo credits: Except as noted, all photos in section 1 are from the Othmer Archives, Chemical Heritage Foundation, and all photos in sections 4 and 6 are of items in the Othmer Archives, Chemical Heritage Foundation, and were taken by Gregory Tobias.

Library of Congress Cataloging-in-Publication Data

Donald Frederick and Mildred Topp Othmer : a commemorative of their lives and legacies / edited by Arnold Thackray and Amy Beth Crow.
p. cm.
ISBN 0-941901-22-X (alk. paper)
1. Othmer, Donald F. (Donald Frederick), 1904– . 2. Othmer, Mildred Topp, 1907– . 3. Chemical engineers—United States—Biography. 4. Chemical engineers' spouses—United States—Biography. I. Thackray, Arnold, 1939– . II. Crow, Amy Beth, 1976– .
TP140.088D66 1999
660′.092—dc21
[B] 99-36436
CIP

♾ The paper used in this publication meets the minimum requirements of the American National Standard for Information Sciences–Permanence of Paper for Printed Library Materials, ANSI Z39.48-1984.

This volume is dedicated to
Don and Mid Othmer,
who, through their great generosity,
have made worlds of possibility
become reality.

Contents

Illustrations

Preface

This book is a tribute to Donald and Mildred Othmer. In grateful recognition of their generosity, we seek to portray their lives and myriad achievements. Theirs was a generosity that took many forms; the diverse selections included in this volume are intended to acknowledge the many communities, organizations, and individuals about whom the Othmers cared deeply.

Don and Mid shared their lives with countless students and friends, and with a widely scattered family. For six decades they were at the center of the chemical community, in New York, the nation, and the world. They took people around the globe with them through the detailed and engaging annual "Dear Folks" letters. They nurtured Polytechnic students, both intellectually and socially. They dedicated themselves to Brooklyn community organizations and gave wholeheartedly of their time, energy, and resources. And they left provisions to ensure that the work to which they gave their great efforts would continue, even in their absence.

No one who knew them well will forget Don's forthright intelligence and practical nature or Mid's quiet wit and flawless taste. Individually they were remarkable people. Together—and "Midon" were always together—they made a formidable combination. Because they chose to devote their lives to *good*, their legacy will endure. And of course, their most immediate legacy is in the lives of Don's (better, their) students. Don wrote in his *Who's Who in America* entry, "The reward to the educator is in the pride in his students' accomplishments. The richness of that reward is the satisfaction in knowing that the frontiers of knowledge have been extended." Through their lives and their generosity the Othmers have thus charged us with the noble task of expanding the frontiers of learning.

Today, the botanist at the Brooklyn Botanic Garden, the doctor at Long Island College Hospital, the first-year student at Polytechnic University, and the scholar at the Chemical Heritage Foundation probably never met Don or Mid Othmer. But through their work and their practical benevolence, the Othmers continue to influence many lives in many places.

Arnold Thackray

Acknowledgments

This volume was made possible in part through the generosity of the Brooklyn Botanic Garden, The Brooklyn Historical Society, the Long Island College Hospital, Planned Parenthood, the Plymouth Church of the Pilgrims, and Polytechnic University. We would also like to thank Warren Buffett, George Bugliarello, Tranda Fischelis, Gerhard Frohlich, Alec Jordan, James Kingsbury, Barbara Kohuth, Mary Seina, and Leslie Shad for their willingness to share personal memories of Don and Mid Othmer.

This volume owes much to the hard work and diligence of many people at the Chemical Heritage Foundation, especially Marge Gapp, Shelley Wilks Geehr, Paul Giblin, Susan Hamson, Frances Coulborn Kohler, and Patricia Wieland.

Their Lives

"Don was just a great guy—a wonderful educator and probably the most widely known chemical engineer in the world."

—*Alec Jordan, founding editor,* Chemical Week

"Mid was a lady of taste, compassion, and understanding."

—*George Bugliarello, chancellor, Polytechnic University*

Don and Mid, pictured with the dogwood tree in front of their house in Coudersport, Pennsylvania, 1983. The Coudersport house, designed by Don in 1938, served as the Othmers' home away from home. They put a great deal of energy into the house, and it was with great sadness that they finally sold it, in the late 1980s.

Section 1

Donald Frederick and Mildred Topp Othmer: A Biographical Essay

by Amy Beth Crow

When Donald and Mildred Othmer decided that they wanted rosewood desks for their home offices, they could have easily left their Brooklyn townhouse, crossed the East River into Manhattan, and bought them. Or they could have chosen desks during one of their many trips to the Far East. Don, the engineer, and Mid, the former fashion buyer for Topp's of Omaha, however, preferred to acquire their furniture somewhat differently. They designed it. Involved with every detail, Don corresponded with the Majestic Company, manufacturers of furniture in Hong Kong, and supervised the construction project, from halfway around the world. Mid, in concert with Don's mechanical drawings, specified the aesthetic touches: the Japanese silk lining of the accompanying wastepaper baskets, the exact hue of the wood, and the decorative carvings ornamenting the pieces. With the woodworking and carving skills of the craftsmen at the Majestic Company thus employed, the Othmers received two beautiful, well-designed, functional pieces of domestic art. They continued to commission various articles of furniture from the Majestic Company, from a silver chest and coffee tables to an impressive quarter-circular desk that Don designed for his office at Polytechnic University* and from which he managed his academic and consulting affairs for thirty-five years.

Desks were not the only subjects of Don Othmer's extracurricular designs. As a child he had planned to be either an architect or an engineer.

* Polytechnic University was founded as the Polytechnic Institute of Brooklyn in 1854. It was known as such until 1973, when it became the Polytechnic Institute of New York. In 1985 the school again changed its name, to Polytechnic University. In this volume, it will be referred to as Polytechnic University (or "Poly"), regardless of date, for clarity.

Mid Topp, as she appeared in her early years.

A youthful Don Othmer. This portrait photograph was inscribed either to his sister, Mildred Othmer Peterson, or to his brother, Kenneth R. Othmer.

His ultimate decision to pursue engineering did not keep him from trying his hand at architecture later in life; during the 1930s he sketched out blueprints for a vacation home in Coudersport, Pennsylvania, after designing a chemical plant in the area. Although he hired an architect to articulate the design more fully, Don provided the vision behind the project and served as the mechanical expert for the many luxuries installed in the eight-room mansion. In 1938 he entered his Coudersport home in a contest for the design of the "New American Home" sponsored by the General Electric Company. Despite his lack of formal architectural training, Don received an honorable mention in 1939 for the meticulous planning and careful construction of his house.

The hands-on approach that both Don and Mid took toward their lives—whether it was in the careful selection of just the right chandelier to complement the foyer or the simple, yet ingenious design of a chemical plant—made them, in the words of longtime friend Warren Buffet, "the perfect partners." From their marriage in 1950 until Don's death in 1995, "Midon," as they were known together, were an international team, affecting people's lives from Brooklyn to Burma.

Born in Omaha, Nebraska, on 11 May 1904, Donald Othmer manifested an investigative nature early on. At the tender age of three, he inferred that the swaying of trees caused the wind. A few years later, when he was older and wiser, Don concluded upon observing the wear patterns of his polka-dotted shirts that the black fabric had been chemically treated to remove color, creating the dots. This treatment, he reasoned, resulted in a more rapid deterioration of the fabric that composed the dots. The geometrically adept Don also deduced the Pythagorean theorem at age ten in an attempt to cross a vacant lot by the shortest distance.

Another Omaha native, Mildred Topp, the second daughter of Holgar and Mattie Topp, was born on 13 September 1907. Mildred's fifth year was full of change: Her younger sister was born, her parents divorced, and her mother bought a yard goods store in Omaha. Mildred, or Mid as she was called, and her sisters, Alice and Nelsie, grew up sharing the responsibilities of housekeeping and helping their mother expand the store's merchandise from bolts of fabric and bib overalls to ready-to-wear clothes. Mid, the most academically inclined of the three sisters, was an insatiable reader and the resident tutor.

Don, the son of a sheet-metal worker, grew up tinkering in his father's shop, where he developed a proclivity for mechanical reckoning, as well as considerable skill in both woodworking and metalworking. Later in his life he would extend this facility to glassblowing. Don was an avid reader from a young age and enjoyed books on the sciences, though he developed no serious inclinations toward chemistry until high school.

Before his senior year in high school Don assumed he would pursue a career that would allow him to use his mathematical and mechanical abilities—some type of engineering. Don never entertained adding the prefix *chemical* to his career intention until he stepped into H. A. Senter's chemistry laboratory at Omaha Central High. Senter, a German chemist, modeled his laboratory on the famed chemistry laboratory of Robert Bunsen; like Bunsen's students half a century earlier, Don and his classmates lit their Bunsen burners from an ever-burning flame in the center of the lab bench. Don was enthralled by his experiences in high school chemistry and decided that his old interest and his new passion had to be combined—he would be a *chemical engineer*!

A scholarship from the Omaha Board of Education enabled Don to enroll in Chicago's Armour Institute of Technology in 1921. Chemical engineering was a nascent discipline in the early twentieth century, and it was widely believed that a background in analytical chemistry was necessary for any chemist in industry; therefore, Armour's program required analytical chemistry. Unfortunately Don detested analytical chemistry, and when he was faced with being compelled to study it the summer after his sophomore year, he decided that this brand of chemical engineering was not for him. He withdrew from the Armour Institute and returned to Nebraska, where he enrolled in the University of Nebraska at Lincoln. Don chose his new school well, for with just a little clever manipulation of the college rules, he was able to complete his bachelor's degree in chemical engineering in one year and avoid any further course work in analytical chemistry. Thus Don graduated from the University of Nebraska in 1924, having successfully condensed his undergraduate education into three years.

A few years later Mid also enrolled at the University of Nebraska at Lincoln, interested, like her future husband, in chemistry. Mid, even in her early twenties, was a woman with broad interests and diverse talents: She majored in both chemistry and economics at the University of Nebraska, went on to teach English at Benson High School in Omaha, and then switched careers, becoming a buyer for her mother's increasingly well-known dress store, Topp's.

After graduation from the University of Nebraska, Don promptly found two jobs: He taught descriptive and solid geometry to members of his father's sheet-metal workers union, and he worked as an analytical chemist

Don Othmer, in his early years.

As elegant denizens of the fashion world of the 1940s, Mildred Topp (left) and her elder sister, Alice, pose in evening wear, circa 1945. At this time Mid was living in New York City and serving as the resident buyer for her family's clothing store in the Midwest.

at the Cudahy Packing Company. (Don recalled in his 1986–87 oral history for the Chemical Heritage Foundation, "I was not doing gravimetric analysis but doing various other types, including nitrogen by the Kjeldahl method and amino acids by the Van Slyke determination. Such semi-instrumentation analyses were not so bad. Besides, the meat-packing industry itself was very interesting.") Neither job lasted long. In 1924 he applied to graduate school at the Carnegie Institute in Pittsburgh. That summer he was boarding at the Alpha Chi Sigma fraternity house at the University of Nebraska, and, as he noted, "A fraternity house in the summertime is rather disorganized, and [the Carnegie offer letter] had been waiting there for me for several days." He went on to recall, "It offered me a fellowship to pay all expenses at the Carnegie Institute. . . . It had been dated ten days earlier; immediately I wired my acceptance. . . . I received a response that unfortunately they [had already filled the fellowship] because it was only a month until school started." Don applied to and was accepted at the University of Michigan. Reflecting on this key moment, Don remarked in reference to the Carnegie offer, "Had this letter been received on time, my education, specialization, and lifework would have been entirely different."

At Michigan, Don completed both his master's degree and Ph.D. in chemical engineering under Walter L. Badger, whom he characterized as a member of the "school of dogmatic, tough-talking professors, bouncing down on every student seemingly at every chance." Badger claimed to have learned his chemistry at the university and his chemical engineering through studying manufacturers' catalogs. Don worked with Badger for three years, concentrating on his mentor's specialty: heat transfer and evaporation.

Although he had planned to follow in Badger's industrial footsteps as an expert in evaporation in the salt, sugar, or chemical equipment industries, a very different career path opened for Don. Upon receiving his doctorate, he was recruited by the Eastman Kodak Company, of Rochester, New York. Don accepted the interview, not seriously considering employment with Eastman Kodak but curious enough to visit the headquarters. Don was impressed with the employees of Eastman Kodak, but Eastman Kodak was even more impressed with the young engineer. Don was immediately offered the job and enticed with an irresistibly large salary.

At Eastman Kodak, Don's first assignment was to develop a means of concentrating acetic acid from its dilute solution. At that time film was a booming industry in more ways than one. Made with cellulose nitrate, a

Don "experimenting" in the chemical engineering lab at Polytechnic University, 1940s.

Throughout his career, Don was always the practical engineer. In this photo the young Professor Othmer inspects equipment in the chemical engineering department at Poly.

compound chemically similar to guncotton, early film was extremely flammable and potentially explosive. In his oral history Don recalled, "Kodak was trying very hard to develop home (16-millimeter) movies in the late 1920s. . . . When projecting the film at home, if the projector happened to go wrong and the sprocket didn't wind, I would just let the film come out in a pile on the floor. . . . That pile of loose film, if of combustible nitrate, obviously could be a very great hazard. . . . [If] somebody dropped a cigarette in it, it would go up!" In the wake of various film fires and explosions—one particularly violent explosion, remembered Othmer, occurred when X-ray film in a storage vault in a Cleveland hospital caught fire—Kodak began a research program aimed at developing safety film.

The contending substitute for cellulose nitrate film was cellulose acetate, a compound that, while flammable, was significantly less volatile than cellulose nitrate. A critical step in the production of cellulose acetate was the concentration of acetic acid through distillation, and this is where Don Othmer stepped in. He set about to learn all he could about acetic acid and cellulose acetate. Combining his mechanical and intellectual skills, he built columns to use in the study of mixtures undergoing the process of distillation; already a capable metalworker and woodworker, Don taught himself glassblowing in order to complete his still. The Othmer still removed water from acetic acid, thus enabling the manufacture of cellulose acetate and safety film.

The invention of the still was Don's first foray into the field of distillation. At Eastman Kodak he furthered his study using azeotropic distillation in an attempt to produce a greater yield of pure acetic acid. His success with acetic acid was noticed: Don was called into various other departments at Kodak to assist with distillation. He went on to separate butanol from water and acetone and methanol—all key chemical materials to the film industry.

Don stayed just five years with Eastman Kodak, leaving in 1931. Although he was the brain behind forty Kodak patents, he received only $10 per patent filed, then watched the company earn millions as a result of implementing the processes he invented or designed. The entrepreneur in Don was dissatisfied with this incentive plan, and he decided to strike out on his own. "I thought to myself, 'Here are millions and millions of dollars going into these plants with processes that I invented and designed.' . . . So I was thinking about getting out to be on my own." He set up shop in the basement of the American Chemical Products Company in Rochester. He continued with his work on distillation and built two new stills almost

Although he never worked in the sugar industry (as he had expected he would, being an expert in distillation), Don did appear on the cover of the January 1949 *Sugar Journal* working with, undoubtedly, distillation equipment.

A diagram of an Othmer still, in an article published by Don in *Industrial and Engineering Chemistry*, in 1928.

immediately. Don's scientific successes during this period were in sharp contrast to the deteriorating economy outside his basement shop. As the Depression worsened, no one was particularly interested in, or capable of, investing in Don's new distillation processes or stills.

Fortunately, Don secured two employment offers in this era of economic gloom: a teaching position at Brooklyn Polytechnic University and an industrial position with Standard Oil of New Jersey. In contemplating his options, he recalled thinking, "I am an entrepreneur, and if I go to work for Standard Oil of New Jersey, I still won't be able to peddle my own inventions." Don foresaw both academic and industrial freedom in a teaching career, plus a fertile opportunity to combine academic research with invention, allowing each to complement the other, thereby increasing the profitable outcome.

In September 1932 the twenty-eight-year-old Don joined the newly independent chemical engineering department at Brooklyn Polytechnic University. Over the course of his long life Don went on to do many things, but he never left the chemical engineering department at Poly nor lost his devotion to the teaching profession and to his students. A year after moving to New York City, Don joined The Chemists' Club to establish connections with the non-academic world of chemical engineering. As the social center for chemists and chemical engineers of the New York area, The Chemists' Club served as a prominent locale from which he expanded his private consulting business.

In his first year in Brooklyn, Don took a pay cut, received a promotion from instructor to assistant professor, and taught Poly's first graduate course in chemical engineering. He also managed to integrate his teaching and research with the industrial consulting that he conducted throughout his career. In his oral history Don could remember only one near disaster occurring as a result of simultaneously teaching and consulting. One winter break he took two of his students with him on a consulting job for Gray Chemical. The job required continuous operation of the pilot plant he had designed for the distillation of ethyl tertiary-amyl ether. Don assigned the night watch to the elder of the students, the day watch to the younger, and he himself split his time between the two. In the middle of the night, in freezing weather, Don got a frantic phone call from the plant telling him that his student had been killed in an accident. As Don recounted,

> I had requested that stairs be built between the platforms around the column. Instead ladders had been built. Bill was checking temperatures, up probably

> twenty feet. Something went wrong, too much steam for some reason or another, possibly the boiler's steam pressure jumped. . . . The column "puked," as we say. Liquid went over the top and out the vent. Too much vapor load and liquid poured out the vents for pressure, and this came down in a tremendous shower. . . . It poured down and showered Bill, who was climbing halfway up.

Fortunately, the story had a happy ending—the student had only passed out from the fumes. Well aware that such risks were a part of the profession, Don continued to let students participate in his extra-academic work, bringing them even as far as Burma. Although the continuous stream of consulting work may have occasionally drawn Don's attention away from his classes, one of his students, Gerhard Frohlich, remembered, "He was an excellent mentor. He gave you the perspective of what a chemical engineer's life in the real world was all about." Don's first doctoral student was granted a Ph.D. in 1937, the same year Don assumed the chair of the chemical engineering department.

The beginning of World War II made Don even busier; since he was past the age for combat duty, his intellectual and mechanical abilities were mobilized instead. Don also found himself teaching increasingly large numbers of army enlistees. As he recalled, "Somebody had told the Department of Defense . . . that college education should be on a three-semester basis. . . . We had to be teaching the same courses to our civilian students on our regular basis of two semesters and to the army—the same courses, the same professors, different students, different clothing . . . on a three semester basis. . . . We had to have a commencement two times for the army and one time for the civilian students." Although he was not aware of it during the time, Don's research was assisting the war effort independently of him: His process for recovering acetic acid, licensed to Eastman, was secretly used in making RDX cyclonites. Later on, Don himself was recruited to "spy school" to train as an evaluator of German chemical plants. He found this experience amusing but was grateful that the end of the war obviated putting this training into practice.

Career opportunities had also drawn Mid to the East Coast during the 1930s and 1940s. As buyers for Topp's, Mid, her sister, and her mother would take frequent trips to New York City. Ever adventurous, Mid decided to remain in New York. She leased an apartment on the East Side and became the resident buyer for the store. She earned a master's degree from Columbia University, and by combining her talent for business with her innate sense of style, she helped transform Topp's into one of the most exclusive and fashionable women's clothing stores in the Midwest.

Mid and Don cutting their wedding cake, November 1950.

Don's postwar career took him around the world. He did not, however, go alone. Somewhere the paths of the businesswoman and the professor crossed. They must have made quite an impression on each other, for in 1949 Mildred Topp brought Donald Othmer home to Nebraska to meet her family over Christmas. The next Thanksgiving they were married in Manhattan. Having agreed early on that they would spend no more than three days apart, Mid and Don became accomplished globetrotters over the course of their forty-five-year marriage. The Othmers shared their enthusiasm for travel with friends and family through a series of travelogues called "Dear Folks" letters. Recipients of these letters were introduced to many exotic locales, as seen through the eyes of the adventurous team, Midon.

Immediately following World War II, Don began work on the project that brought him international recognition. In the early 1940s he and Raymond Kirk, his counterpart in Poly's chemistry department, decided that an encyclopedia of chemical technology comparable to Ullmann's *Encyclopedia* was needed. The first volume of their encyclopedia was published in 1947, covering the top of a list of more than one thousand topics they had laid out for the series. Today the *Encyclopedia* is in its fourth edition and comprises twenty-seven volumes. It is found in virtually every university, research facility, and company that employs chemical processes. It is an invaluable resource for chemists and chemical engineers everywhere.

Also in the 1940s Don's years of distinguished work in distillation prompted an offer from Steel Brothers Limited of Great Britain to design a plant for the distillation of kerosene in Burma. Although he never visited Burma during this project, another opportunity to do so surfaced soon after Don and Mid's wedding. Don undertook a major consulting project with the international engineering firm TAMS, a company extensively involved with the expanding Burmese chemical industry. So valued was his expertise that TAMS wanted to relocate Don and his wife to Burma, where they would be settled in a colonial mansion and outfitted with eight servants. Don refused the offer, citing professional responsibilities and dedication to his students. He agreed, however, to design plants from his office in Brooklyn and to send his top Poly student to oversee construction of the Burmese plants. Of course, Don and Mid also paid numerous visits to Rangoon.

Always eager to explore new challenges for chemical engineering, in the early 1960s, Don began working on various desalination processes with the U.S. Department of Commerce's Office of Saline Water. In his

Don and Mid pose in full Japanese garb in Tokyo, August 1952. This photo was taken for their Christmas card of 1952. The card's greeting was indicative of their globetrotting nature: "Wherever we are—Christmas '51, Valparaiso, Chile; Christmas '52, Rangoon, Burma; and However we are—We send *You Cheers!* for Christmas and for *1953*."

oral history Don recalled that his interest in desalination originated during his work as a doctoral student in Badger's evaporator laboratory at the University of Michigan. In his thesis research Don had developed a new system of calculations for designing multiple-flash-evaporation desalination plants. Characteristically, Don carefully studied the existing process of multiple-flash evaporation and then improved upon it. Commercial desalination plants had traditionally used a process in which salt water was converted to a vapor and then distilled via controlled changes in atmospheric pressure. Don replaced the traditional tube condensers with spray condensers, thus allowing an open expansion of the system. Don called his new process the "vapor reheat system" and successfully licensed several patents related to it.

In a similar vein Don became interested in exploiting differences in ocean temperatures as an alternative energy source for developing countries. He knew that there was a great deal of thermal energy present in the temperature gradient between the very cold, deep water of tropical oceans and their warm surface layers. In the late 1960s Don designed a small plant in the Bahamas to produce both fresh water and electric power. Later he improved this system so that it also collected the nutrients present in deep water for use in the cultivation of shellfish. These efforts all reflected Don's growing concern about the rapid exhaustion of the planet's natural resources. He saw an urgent need to develop processes that made efficient use of renewable resources. As he noted in a paper delivered at the Center for the History of Chemistry at the University of Pennsylvania in 1987, "This explosion of science, technology, and industry has been at the expense of a vast amount of the world's resources." Don's activities in this area were a professional expression of his community-oriented nature.

The 1970s and 1980s saw Don and Mid busy with their international travels. Their Dear Folks letters took readers from Yugoslavia to Kuwait, from Germany and France to Saudi Arabia, Iraq, and India. Don's pioneering work in petroleum processes and water resource management brought scores of invitations to conferences around the world. The early months of 1976 took the Othmers to Saudi Arabia, and they opened their first Dear Folks letter of that year, "Just four weeks ago today, we embarked on a magic carpet directly to Arabia, a journey that took part of three days and touched on three continents." The second half of that same year, however, "took us to no exotic places unexplored by us previously," thus the Othmers settled for a description of their visits to "London, Frankfurt, and Lisbon—all places that, because we have visited them frequently in the past, seem like old friends."

As intellectually and inventively active as ever, Don wrote in 1972 to longtime friend and associate Robert Lyman:

> Last night, till 2 A.M. I was writing a proposal for an experimental plant to make electric power and fresh water using as a source of energy the difference of temperature, 40–45° F, between surface and deep-sea water in tropic seas.
>
> The night before it was systems for extracting aluminum from domestic ores rather than imported bauxite. I am working on, writing papers and

The September 1977 issue of *The Chemist*, the magazine of the American Institute of Chemists, featured Don and his most famous contribution to the scientific community: the *Kirk-Othmer Encyclopedia of Chemical Technology.*

patents on a half dozen fields right now. I did give up that cancer research effort a year ago.

The point is I get into these fields of interest—very tangible fields due to one reason or another, and they keep me so busy that for weeks I have hardly had an hour, Saturdays, Sundays, or nights to get away from my desks (office and home) to do the several jobs of wood, metal, or plastic, or electrical work which I do enjoy for relaxation.

Advancing years certainly did not slow Don Othmer down; if anything, his interests expanded with time. His foray into cancer research, though brief, shows the remarkable range of his pursuits.

Longtime residents of Brooklyn, Don and Mid gave generously of their time and money to their community. Together they supported the Long Island College Hospital, where Don was elected to the board of regents in 1968. A year later Mid joined the Women's Guild of the hospital. The Othmers also donated money for the construction of a center for cancer research at the hospital—the Donald F. and Mildred Topp Othmer Building. Mid pursued her own interests as a volunteer for the Brooklyn Botanic Garden and as an ardent supporter and board member of Planned Parenthood. The pair devoted much of themselves to Polytechnic University: Don, as a mentor and friend to many students, and Mid, as the impeccable hostess, ensuring that all visitors—associates, students, or friends—were well cared for in the Othmer household at 140 Columbia Heights.

Through the course of his long career Don accumulated some 150 patents and saw his work applied to fields as diverse as resins, plastics, pharmaceuticals, and waste treatment. He also earned many prestigious awards, including the 1978 Perkin Medal from the Society of Chemical Industry (American Section) and the 1991 Founders Award from the American Institute of Chemical Engineers. The American Chemical Society has an Othmer Building at its headquarters in Washington, D.C., and the American Institute of Chemical Engineers administers a number of student scholarship awards in Othmer's name. As an educator, Don sent many successful chemical engineers out into the world. Although he was made professor emeritus in 1976, he never officially retired from Poly and was actively involved with the school until his death in 1995. When asked to reflect on his lifetime of experiences, Don said, "I have always enjoyed what I did. . . . [E]verything I have had to do I've taken in stride and enjoyed it."

From Omaha to Hong Kong the Othmers had friends around the world and were guests as frequently as they were hosts. Don and Mid continued to be dedicated participants in and contributors to their communities and causes to their final years, and even thereafter. When Mid died in 1998, the depth of Don and Mid Othmer's dedication and support was made astonishingly apparent. Having invested wisely, and with characteristic patience and practicality, the Othmers had amassed a combined estate worth $800 million. And in a gesture of exceptional philanthropy, they gave this wealth away to the institutions and organizations that had enriched their lives and that they had so cared for. Both Don and Mid Othmer lived full and complete lives without the loud trappings or financial preoccupations that may accompany great wealth; the story of their lives is deepened, but not defined, by their altruistic decision to leave their fortune to their favorite charities.

The Othmers will be justly remembered for this act of uncommon generosity. However, their lives were infinitely more than the sum total of their estate. Don's many scientific achievements changed the face of chemical engineering in the twentieth century, and his legacy will live on through the accomplishments of his students. Likewise, Mid remains the paragon of graciousness and refined taste in the hearts of those who knew her. Undoubtedly she will also be remembered for the indelible mark she made on the fashion community and on the many charitable institutions to which she gave selflessly of her time.

Section 2

The Othmers in the *New York Times*

OBITUARIES

Friday, 3 November 1995

Donald F. Othmer Dies at 91; Innovative Chemical Engineer

by Robert McG. Thomas, Jr.

Donald F. Othmer, an innovative chemical engineer who combined an inquisitive mind, a practical nature and gift for manual dexterity into an acclaimed career as an educator, inventor, entrepreneur and philanthropist, died on Wednesday at Long Island College Hospital in Brooklyn. Dr. Othmer, who had been a professor at Polytechnic University for more than 60 years, was 91.

Although Polytechnic was his base from the time he joined the faculty of the university's fledgling chemical engineering department in 1932, Dr. Othmer, who served for many years as department chairman and helped transform the obscure college in Brooklyn into one of the nation's leading technological centers, had a wide-ranging career.

Among other things, he helped develop one of the most powerful conventional explosives, RDX, which was made from acetic acid and used in World War II. He also obtained more than 150 American and foreign patents, and circumnavigated the globe more than a dozen times as a highly sought international consultant who virtually created the Burmese chemical industry after the war.

As well known to generations of chemical engineers as Hoyle is to card players, Dr. Othmer was a founding editor of the *Kirk-Othmer Encyclopedia of Chemical Technology*, a multivolume reference work now in its fourth edition. (The co-founder, Raymond Kirk, a former chairman of Polytechnic's chemistry department, died before the first edition was published in 1947.)

A native of Omaha, where his father was a sheet-metal worker, Dr. Othmer traced his interest in chemistry to a high school teacher who had studied at the University of Heidelberg in Germany with Robert Bunsen, whose Bunsen burner became a laboratory fixture.

After graduating from the University of Nebraska and receiving a doctorate from the University of Michigan in 1927, Dr. Othmer assured his own fame as a young engineer for the Eastman Kodak Company.

Assigned to find an efficient way to distill acetic acid, a major ingredient in safety film, he not only accomplished that task, laying the groundwork for a huge new industry, but also did it in a particularly ingenious way.

To measure the effects of various distilling methods, he invented the Othmer still, a device that continues to be used to study distillation.

It was typical of Dr. Othmer's hands-on, low-cost approach to scientific problem solving in that he not only dreamed up and designed the still, but also learned glassblowing so he could build it himself.

After receiving a $10 bonus for each of the 40 or so patents *he* obtained for Eastman, some eventually worth millions to the company, Dr. Othmer struck out on his own, working as a consultant until the Depression led him to the security of a post at Polytechnic—an institution, he later noted, that assured him both academic and entrepreneurial freedom.

Dr. Othmer, who had a knack for finding simple, practical solutions to complicated problems, specialized in improving chemical manufacturing processes. His work has been fundamental to the production of billions of pounds of resins, plastics, surface coatings, textile fibers, foods and pharmaceuticals.

Not all of his chemical inventions paid off. Most tended to be elegant, however, although few reached the inspired Rube Goldberg heights as the 1975 patent he received for a process to use sunlight and seawater to produce fresh water, electric power and nourishment for clams, oysters, lobsters and other seafood.

Dr. Othmer gave away millions of dollars to charities, but the money did not come primarily from his patent royalties. Rather, it came from his investments. As skilled in finance as he was in nearly everything else, Dr. Othmer, a man of such practical frugality that he dispensed with lawyers and wrote his own patent applications in longhand, was one of the earliest partners of his Omaha friend, Warren Buffet.

He is survived by his wife, the former Mildred Topp of Omaha, who now lives in a nursing home there, and a sister, Mildred Peterson of Chicago.

Monday, 13 July 1998

Staggering Bequests by Unassuming Couple

by Karen W. Arenson

In many ways, Donald and Mildred Othmer were like any other faculty couple. He was a professor of chemical engineering at Polytechnic University in Brooklyn, a workaholic with scores of patents and a sideline in consulting. She was a former teacher and a buyer for her mother's dress stores who volunteered at the Brooklyn Botanic Garden and Planned Parenthood. They had no children, unless you count the students he invited home to dinner.

When they died—he in 1995 and she in April, both in their 90s—they left their money to many of their favorite charities, including his university and the places where she volunteered.

But the level of their generosity distinguished them from the typical academic couple, putting them more in a league with the Ted Turners and George Soroses of the world. The value of their combined estates is approximately three-quarters of a billion dollars.

Although Dr. Othmer was a successful inventor and a tireless consultant, there is another explanation for the stunning total. Some decades ago, they invested most of their savings with an old family friend from Omaha: Warren E. Buffett, the stock market wizard who runs Berkshire Hathaway, the investment and insurance holding company. And thanks to patience, the booming market and Mr. Buffett's golden touch, the Othmers were transformed over the years from comfortable to wealthy to staggeringly rich—all with barely an outward sign of change.

Now it is the beneficiaries of their gifts, some of them modest institutions of little renown, that stand to be transformed, with the Othmers' bequests instantly multiplying their resources and vastly broadening their possibilities.

Polytechnic, for instance, stands to receive nearly $200 million, a sum about four times the school's entire endowment. Long Island College Hospital, also in Brooklyn, is in line for about $160 million. The University of Nebraska and the Chemical Heritage Foundation, in Philadelphia, also expect to get more than $100 million each, while a number of other institutions will also get multimillion-dollar gifts.

"It is an amazing estate," said Theodore R. Wagner, the New York estate tax lawyer who helped the Othmers draw up their wills. "It is really quite astounding that this nice couple was able to drop so much money on so many institutions."

For the recipients, the gifts are like winning the lottery. Officials at Polytechnic see the money as a way for the university of 2,000 students to climb from a second- or third-tier engineering institution into more selective company, with an endowment per student larger than that found at Rensselaer Polytechnic Institute, in Troy, N.Y., or Carnegie-Mellon University, in Pittsburgh. "We start from being one of the have-nots and go to being one of the very well-endowed schools," said Polytechnic's president, David C. Chang.

The final amounts of the bequests will depend both on legal negotiations surrounding the wills and on the stock market.

One set of negotiations is between the charitable institutions and Mary Donahoo Seina of Omaha, Mrs. Othmer's niece, who is to receive less than $2 million under Mrs. Othmer's will. Mrs. Seina, who is in her mid-60s, declined to be interviewed, said her lawyer, James Fitzgerald. But she told the *Omaha World-Herald* in May that her own family deserved more, and that her aunt no longer wanted to leave her money to the same institutions and had revoked her will. Mr. Fitzgerald said only that the settlement talks were continuing.

There are also questions about how much of the estate should be disbursed under the terms of Dr. Othmer's will and how much under Mrs. Othmer's; that issue could affect exactly how much each institution receives.

And, of course, the sums will also be tied to the value of Berkshire Hathaway shares. After initially investing $25,000 each in a partnership with Mr. Buffett in the early 1960s, the couple received shares of Berkshire Hathaway in 1970 at $42 a share. The stock closed on Friday at $77,250 a share.

Lawyers say that Mrs. Othmer's estate contains about 7,500 shares of Berkshire Hathaway stock; her husband's approximately 7,000 shares were sold after his death in 1995, at just under $30,000 a share.

While the huge gifts hold the promise of substantial change for some of the beneficiaries, the enormous wealth had little effect on the Othmers themselves, who lived comfortably but not ostentatiously, and rarely talked about their money.

W. Fred Schurig, a protege and colleague of Dr. Othmer, recalls Dr. Othmer's making only one brief reference to his investments, in the elevator at Polytechnic, decades ago.

"He came back from Omaha and was telling me about this guy Buffett and how investing with him was going to be lucrative," Dr. Schurig said.

"I said I had $5,000 and asked, 'Would he take that?' But he said, 'That's not enough; he wouldn't touch that.' "

"He never brought it up again," Dr. Schurig said.

The shares made the Othmers rich many years ago, and during their lives they had quietly made a number of multimillion-dollar gifts. But it was only in recent years that their wealth shot into the stratosphere. At the time the Othmers wrote their wills in 1988—Dr. Othmer drafted much of the meticulous detail himself, down to a small Rembrandt etching and 12 carved elephants, each in a different Burmese wood—Berkshire Hathaway was selling at about $4,700, and their shares were worth about $68 million. (He revised his will in 1994.)

Even without the Buffett millions, it is easy to see Dr. Othmer as the classic American success story: a poor boy who reached the top of his profession, widely known as a consultant and co-editor of the *Kirk-Othmer Encyclopedia of Chemical Technology*.

Born and raised in Omaha, he developed a lifelong frugality as he earned money picking dandelions from neighbors' lawns, delivering newspapers and telegrams and walking a farmer's cow to and from pasture. After graduating from the University of Nebraska and receiving a Ph.D. from the University of Michigan in 1927, he set out for Rochester and Eastman Kodak.

His research contributed to more than 40 patents for Kodak, most of them involving the distillation and concentration of acetic acid. But unhappy to be making only a $10 bonus for each patent, he headed out on his own in 1931.

It was the Depression, however, and business was slow, so when Brooklyn Polytechnic made him an offer in 1932, he accepted. Polytechnic gave him a lab, allowed him to collect his patent earnings and provided graduate assistants for his research and consulting. Unless he was traveling, he was usually in his lab six days a week.

By conventional measures, he was not much of a teacher. Former students recall that he was so wrapped up in research and consulting that he spent little time preparing lectures and would abandon class to take calls from his consulting clients. But to generations of ambitious young students, many of them first-generation immigrants, he was the ideal role model and mentor, one who involved them in his research, took them home for dinner and found them jobs. Some named their children after him. Others, like Joseph J. Jacobs, called him the most important influence in their lives.

"He was not only an excellent chemical engineer, but he was also ambition incarnate," Dr. Jacobs, chairman of his own engineering firm in Pasadena, Calif., wrote in his autobiography, *The Anatomy of an Entrepreneur.*

Dr. Othmer's first marriage fizzled. But then he met Mildred Topp, known as Mid, a former high school teacher who received a master's degree from Columbia Teachers College in 1945 and was a buyer for her mother's fashion shops in Omaha. Friends describe her as Dr. Othmer's intellectual equal, someone who shared his devotion to his career but was also able to stand up to him.

"Mid was very smart and a full participant in decision-making," Mr. Buffett said in a recent interview by telephone.

They were married at the Plaza Hotel in 1950, and settled into a town house in Brooklyn Heights with a view of the harbor stretching from the Statue of Liberty to the Brooklyn Bridge. They lived on two floors and rented the other three.

Mrs. Othmer was an active volunteer while her husband distilled chemicals. She helped out in the Botanic Garden's plant sales and at Planned Parenthood's thrift shop. She was also on Planned Parenthood's board, while her husband became a trustee of the Long Island College Hospital, just a few blocks from their house.

Dr. Othmer, a tall, solid man with wavy brown hair, impressed everyone with his insatiable curiosity and his strong will. When other professors planned a new encyclopedia on chemistry, for example, he muscled his way into the project, convinced that it should include chemical engineering to broaden its interest to industry. He became co-editor of the resulting *Kirk-Othmer Encyclopedia of Chemical Technology*, which quickly became the industry bible.

At the Othmer home, Dr. Othmer's students, colleagues and clients were frequent guests. So were visitors from Omaha, including Mr. Buffett, who became a good friend.

As Mr. Buffett recalls it, Mrs. Othmer's mother, Mattie Topp, first approached him about investing for the family around 1958, when he was 27 and managing less than $1 million. She had used Mr. Buffett's father as her broker before he became a United States representative.

Other family members withdrew money, but the Othmers were patient investors, adding money from time to time. When Mr. Buffett dissolved the partnership in 1969 and the Othmers took shares in Berkshire Hathaway, Dr. Othmer's investment had grown to $770,000 and his wife's to $817,000. "They just rode along," Mr. Buffett said, adding that the growing investment "never changed their lives."

It was only in recent years that some of the Othmers' intended beneficiaries began to get a clue about the windfall they had coming. Even those who knew they would inherit had little idea of how much.

"Don and Mid had said, 'Don't worry, we're putting you in our wills,'" said Arnold Thackray, president of the Chemical Heritage Foundation, a group dedicated to the history of chemistry and chemical technology. "When someone does that, you smile and say, 'Thank you, that's very nice.'" But Mr. Thackray never expected that the foundation would receive more than $100 million.

Mr. Buffett is used to dealing with large sums of money, but he, too, calls the Othmer bequests an "amazing story."

"They were such high-quality, nice people, who had no children and wanted to translate their wealth into something beneficial to society," he said.

And Mr. Buffett suggested, somewhat tantalizingly, that there might be others whose investments with him will eventually yield vast bequests. "There are more coming," he said. "There are going to be some bigger ones than this."

Op-Ed

Friday, 17 July 1998

How Millionaires Get That Way

by David Frum

The millionaires next door? To me, Donald and Mildred Othmer—the "unassuming" Brooklyn couple who recently left a staggering $800 million estate to charity—were the millionaires in the basement.

In the week since reports were published about the immense fortune accumulated by this onetime Brooklyn Polytechnic professor and his wife, the Othmers have become paragons of thrift: save your money, invest carefully, and you, too, can become rich. And that's true, so far as it goes. But thrift is a virtue more comfortably praised than practiced.

In the spring of 1989, my wife and I arrived in New York looking for a place to live. Those were the last days of the 1980s real estate boom, and it quickly became apparent that we could not afford to live in Manhattan.

So, a week or so later, we found ourselves in the Othmers' parlor, drinking sherry and discussing the rental of the third floor of their brownstone in Brooklyn Heights.

It was a magical apartment, with a balcony overlooking New York Harbor, a great old fireplace and a six-foot-long bathtub. The house had been built in 1840, and a century and a half later still boasted its original doors of Honduran mahogany.

Nothing vanishes so irretrievably as the manners of another age, but the Othmers had somehow preserved intact the domestic habits of respectable society in their native Omaha during the first decades of this century. Mrs. Othmer spoke in the way that elocutionists used to teach, with round vowels and gentle tones. We came to terms without lingering a second longer than necessary on the distasteful subject of money, and then spent the rest of the hour talking about Edith Wharton.

A few months after we moved in, the Othmers invited us to dinner. This became a regular kindness, and each time they apologized for not having us more often.

They practiced a very old-fashioned style of hospitality: highballs and canapés of sliced American cheese on white bread before dinner, a brass bell at Mrs. Othmer's elbow to summon the first course, sauces made from cream of mushroom soup, water with the meal, port and Madeira afterward.

They showed us their collection of Japanese fans and swords, bought at flea markets while Dr. Othmer was advising the Occupation forces after World War II.

I once asked them whether they knew their fellow Nebraskan, Warren Buffett. Mrs. Othmer replied that her mother had known Mr. Buffett's father and had become one of the junior Buffett's first clients when he went into business for himself. "Mother," she said, "always believed in encouraging young people."

The Othmers were clearly rich: The newly restored Brooklyn Historical Society had a big plaque thanking them in its lobby. But it was also clear that they knew the value of a dollar in a way that our softer generation does not. As our money slipped through our fingers, as money does in New York, we often found ourselves wishing we could be more like the Othmers. But we never quite pulled it off. Nor did we ever quite adjust to being on the receiving end of their frugality.

Although crippled by a stroke, Dr. Othmer insisted on personally examining every household problem before permitting a repair service to be called in. He would put his special handyman's baseball cap on his handsome, tall head, gather his toolbox, painfully haul himself up the stairs, reminisce for a while about his wartime work distilling gasoline out of banana skins in Central America, gaze at the problem, tinker with it and then gaze at us reproachfully.

He would point to our toaster oven and comment gloomily: "That's the biggest toaster I've ever seen. Do you know a toaster oven uses as much power as a motorboat?" (I still have no idea whether this is true.) Dr. Othmer cooled his own body with a wheezing electric fan. He was prepared to concede that not everybody could live up to his standards of self-denial. But he was scandalized that we had bought the second-largest window air-conditioner sold at Macy's. "What do you need such a big air-conditioner for? You'll blow out every light in the place."

He was right about that. The Othmers' house had undergone a major renovation during the Wilson Administration, when it acquired electricity and running water, and had been left largely untouched since then. In the mornings, the hot water lasted about 30 seconds. Never mind the air-conditioner: There was not enough power to make toast and coffee at the same time. The wires leading to the fuse box on our floor were wrapped in an antique insulating cloth, and if you were not careful as you tried to get the lights back on, sparks would shoot out, landing disconcertingly close to the woodwork and dried paint of the hallway.

Every once in a while, and especially once our first baby was on the way, I'd remonstrate with Dr. Othmer about the sparks—invariably without success.

Even in his frailty, he was a man of decided views. The front doorbell rang one afternoon when I happened to be home writing. I sprinted for it, knowing how painful it was for Dr. Othmer to drag himself out of bed, but he managed to reach it only a minute or so after I did. There was a shaggy young man canvassing for Greenpeace. I was politely telling him "no thanks" when Dr. Othmer shuffled up behind me.

"Greenpeace?" he barked. "Aren't you the folks who blew up that French boat in Australia?"

He was referring to the *Rainbow Warrior*, the Greenpeace vessel blown up by French intelligence lest it interfere with a Pacific nuclear test.

The canvasser stammered in amazement. Dr. Othmer slammed the door. As he headed back to his apartment, I could see him worrying that his memory had betrayed him

"It was Australia, wasn't it?" he asked me.

"New Zealand."

"That's right," he said with relief, satisfied that he had remembered the essentials of the story.

The Othmers have been hailed as representatives of what is best in America. They were indeed fascinating, cultured people, who carried into the age of Ivan Boesky the hardy frugality of the pioneers. They epitomized virtues that sometimes seem to have vanished altogether.

But looking back on those two curious years upstairs from them, I must say that these virtues are best appreciated from a bit of a distance.

David Frum, a senior fellow at the Manhattan Institute, is writing a book about America in the 1970s.

Section 3

Colleagues, Friends, and Family Remember Don and Mid

W. Alec Jordan

Most biographical writings can be readily categorized: the statistical type that cites this-then-and-thats with a sequence of dates; those dominated by one aspect (e.g., size of estate and bequests); or the substantive essay. All are included in this memorial book.

This piece, however, is unlike any of those. It is simply my personal recollections of a cherished friendship that spanned more than forty years. I hope my anecdotes and observations portray a few facets of the characters and personalities of two truly remarkable people.

In those registers in which we are all recorded, they were, properly, Donald Frederick and Mildred Topp Othmer, but I never heard anyone refer to them other than as Don or Mid. (An occasional graduate student might venture a mumbled "Doc," which subsequently could be easily transmuted to Don.) To my mind that is significant: It reflects the warmth of their characters and some true Nebraskan virtues—friendliness, accessibility, unpretentiousness. (To be sure, they were both diligent and prudent as well.)

Don was exuberantly gregarious and a spontaneous raconteur. He could dominate any meeting or platform. He reveled in the monthly meetings of the Société de Chimie Industrielle and was widely regarded as the doyen of The Chemists' Club, where he held a life membership. In contrast, Mid was quiet, almost shy, but thoughtful and a careful listener. Her observations and judgments were precise and unwavering.

One of my treasured memories is of seeing them walking together. The tall, broad-shouldered Don, leaning slightly forward while gently supporting the elbow of petite Mid. It reflected their devotion, as did the pact they made when they were married. To wit, if Don had to be away more than three days, Mid would accompany him on the trip.

One result of this pact: Because Don lectured worldwide and was a consultant to corporations and governments in some fifty countries, they took fifteen to eighteen round-the-world trips together. (Don was not sure of the exact number.) These forays were pre-jet age when international air travel was slower, more difficult, arduous, and restricted.

Don said that they always circled the world from East to West. You see, Don was of Scottish extraction and Mid had Danish antecedents. It is widely known that the Danes love food and the Scots love a "bargain." So as Don told me, whilst Mid's eyes sparkled, "You pick up an extra day that way . . . three extra meals!"

Mid and Don had another form of partnership, too. They occupied two stories of their five-story house on the Heights in Brooklyn that overlooks lower Manhattan with a magnificent view of the Statue of Liberty. They had tenants in the upper three floors. Mid looked after all the business aspects because, as Don said, "she is better at it than I am." But Don, because help was hard to find and because he said he was "sort of handy," was THE maintenance man. Not too long ago, when he was well into his eighties, I saw Don lug a fifty-pound pail of tar up to the roof to patch a leak. He spilled some, and it secretly pleased me a bit to see this small indication that he was not infallible.

In recent years Don's accomplishments have become more widely known. However let me take you back almost fifty years. At that time I was an editor. It was the week between Christmas and New Year's—about the only week in the whole year that editors have a little time to breathe. So my colleagues and I were relaxing, chatting, laying out plans for the following year. One of my colleagues was the late, renowned editor of *Chemical Engineering*, president of the AIChE, and SCI Medalist, Sidney Dale Kirkpatrick. For some reason I have since forgotten, Sid began to talk about Don. "He has done more," he said, "than *anyone* to spread knowledge of chemical engineering to the far corners of the world. . . . He is a real giant."

What Sid said those decades ago impressed me then. What has impressed me more since is that I have noticed that very word recurring time and time again—*The Chemist* called Don a giant in innovation; *Chemical &*

Engineering News called him a giant of industrial chemistry; the SCI called him an international giant.

A pleasant phrase often used in referring to great achievers is that they "are legends in their own time." Often that is untrue; most legends are generated by history. However, Don obtained legendary status when he was barely forty years old, and, to be sure, he maintained this status through the subsequent five decades of his life. What he accomplished is nothing short of gigantic—engineer, author, editor, inventor, consultant, world traveler, educator, philanthropist. It is almost impossible to believe that any one man, on his own, without a large staff, could achieve so much.

Actually, no part of our daily living is untouched by Othmer-developed technologies—the gasoline that powers our automobiles; the fuel that heats our homes; the paper used to produce newspapers, magazines, and this book; a whole range of plastics and synthetic fibers; a spectrum of antibiotics; and even aspirin.

Mid, too, was multitalented and diligent. She was a recognized dress designer, a professional buyer of ladies wear for the Topp-owned stores in Omaha, a gourmet cook, and a hard-working supporter of her church and a number of local charities. I remember calling Don one snowbound day in February at 8:00 A.M. He was at his office and mentioned casually that Mid had gone to New York to work for one of her charities.

Don't conclude, however, that they were drudges, that is, burdened with seven dull virtues. Far from it. Don, a vigorous, zestful man, had workshops at his homes in Brooklyn and the mountains of Pennsylvania. He designed and made furniture with the finesse of a cabinetmaker, had a huge collection of antique maps, planted thousands of trees on his secluded five hundred acres, and chopped down huge trees—by himself. Mid, of course, could indulge in her creative cooking at either place, much to Don's delight as well as the delight of the many students from abroad who partook of their warm hospitality.

Let me add that Don was an intense conversationalist. When he was talking to you, he *looked* at you—intently. To accompany him when he drove to Manhattan was an unforgettable adventure. Don never hesitated to carry on a conversation while he drove, and of course, if he was talking with you, he was looking at you—glancing now and then at traffic. Those trips, I believe, yielded my first strands of gray hair.

Mid and Don were both proudest, I'm sure, of Don's students who are now scattered all around the world in industry, government, and academe. When Don's students talk of their studies under him, they always dwell on how they benefited from his knowledge and worldly wisdom. And all talk about his friendly encouragement, his warmth, his understanding. Over many years, former students organized birthday celebrations. On his seventieth, as I recall, they commissioned an oil portrait for "Doc."

There is another facet of Don's activities about which very few people know anything. It is a parallel to the old adage, "The easiest person to sell anything to is another salesman." What I know to be true is that the easiest person to sell an invention to is another inventor. All inventors are by nature optimistic. Many have "sugar plum" delusions, but at the least they have confidence and enthusiasm. All of which yields an innate compatibility of interests.

Consequently, Don, apart from selling or licensing his own patents, was constantly assailed by other inventors and entrepreneurs with "great ideas." His resistance level was low; he supported many projects.

One proposal I recall was an idea of the famous engineer J. V. N. Dorr. He envisioned the creation of a partnership to manufacture construction materials (e.g., building board) from then-waste materials. (I don't recall why, but I opted out of that one.) My dim recollection is that project didn't fly.

Then there was the chemist who claimed he could make silk from chicken feathers because of their chemical kinship. (Anyone want to guess how many *tons* of feathers are produced daily by poultry processors?) There was also the chap who was developing "electronic anesthesia": Turn the switch and the patient is out; turn it back and the patient is conscious. (The problem—according to some Russian research—was that the recovered experimental animals sometimes remained a little, shall I say, off-beam.)

However, there was one development that intrigued both Don and me, as a result of an unusual coincidence. In my first job as a research chemist, one of my colleagues was an elderly man who, in his youth, had been Charles M. Hall's lab boy. He regaled us all with tales about this illustrious inventor of the basic process to make aluminum. Years later, when I was working in New York, several people, over several months, brought rumors to me of a new process to make aluminum being developed by an engineer in Louisiana. Naturally I was intrigued and asked my bank to track it down.

Subsequently, when I was having lunch with Don, he mentioned he had just had an interesting trip to Louisiana: He visited a former student who had called Don to seek his advice about a new way to make aluminum. The three of us met in New York and again in Louisiana. The chemistry of thc process looked very good. It was environmentally superior—no waste; the economics appeared to be superior—it could operate on domestic clays rather than imported bauxite. However, an unforeseen shortcoming of the aluminum product prevented full-scale commercialization at that time.

In his later years Don was working on an immense project to produce vast quantities of energy. The concept was, as far as could be foreseen, doable. However, the problems and politics of setting up supply and financial resources were daunting. The significance, to my mind, is that even as he approached ninety, Don was still thinking of, and planning, as always, the gigantic. Perhaps a decade or two hence, this vision will become a robust reality and yet another Othmer technological legacy. Nonetheless, even if it were to become hugely beneficial, it could never rank with Don's and Mid's lifelong—quiet and private—generosities nor with the scope and thoughtfulness of their lives and bequests.

Arnold Thackray

(Presented at Donald F. Othmer's Memorial Service, 4 December 1995)

We are met here to celebrate the life and achievement of a chemical engineer—a consummate engineer, who embodied the ideals of his profession. But we each mourn the losses in our own lives which come with the death of this acute, driving, perspicacious, and multitalented man.

May I quickly remind you of seven facets of Don's life and personality?

First and foremost I remember his compelling presence. It was not just that he was a tall and handsome man, though that was certainly so. It was rather that his robust, clear, and powerful intellect found expression in a particular combination of simple command and quiet humor that brooked no argument. I remember vividly the day on which I first saw him—at more than ninety years—admitting to illness and old age. Arriving at his comfortable, convenient, and, above all, functional home at 140 Columbia Heights, I found Don seated by the dining room window enjoying his unsurpassed view of New York harbor (what common sense to find and buy that quiet house, not only close to subway, church, and school but "the nearest house to Wall Street" at a time when the suburbs were in fashion and urban prices were low). I asked him how he was. "It's not a day on which I'd volunteer for a calculus test," he immediately replied. In twelve words he'd reminded me that he was an academic, an engineer, and in command of his faculties and his fate, however precarious his circumstance. Equally compelling in its many implicit messages was his oft-given terse advice: "Buy a good stock and hold it for fifty years."

Second, I remember his love of correctness and control. Don always had the right word—*le mot juste*—and hence was a great editor and a great teacher. I know that Don's students and colleagues in chemical engineering, from Poly, from *Kirk-Othmer*, and from around the globe could tell many tales of his remorseless editorial zeal. Let me only say that in the last seven or eight years of his life, Don could no longer devote himself fully to research and writing. However, the discussions that led up to the creation of the Othmer Library necessitated that many ideas be explored in writing. I quickly learned that even my proudest professional efforts should be labeled "draft," for Don took enormous pride and pleasure in editing and (I admit) improving every piece of prose I ever laid before him.

The third thing I shall remember is Don's essential modesty and practicality. I think of him agreeing to speak at an after-dinner meeting in Philadelphia. "Could I pick you up at the station?" I asked. "No, it isn't necessary," he replied. And so at the end of a working day at Poly, the eighty-three-year-old Don voyaged by subway, train, and on foot to the Chemical Heritage Foundation for dinner, a talk, and back to bed in Brooklyn. No need to fuss! The logistics, the engineering of the situation, were obvious to anyone who made a straightforward analysis.

Fourth, I think of Don's probity. One of many examples occurred when I first visited his office on the fifth floor at Poly. I stepped out of the elevator. The floor was deserted, but in the distance I could hear a telephone ringing in what was obviously Don's office, the door ajar. Just then, Don himself appeared around a corner. Don, of course, lived in the world a long time before the advent of voice-mail and answering machines. "Don," I said urgently, "your phone is ringing." "Of course," said Don. "See, I have two separate lines—one is Poly's, which is for my academic work. The other is for my consulting, and I pay for it myself. So when I leave my office for a minute, I call one phone from the other. That way, callers on both lines hear a busy signal and know to call again in a little while."

This story illustrates a fifth fundamental trait of Don's—his preeminent practicality. I think of his bedside lamps, which he machined from shell cases in such a way that they could be dimmed by a simple turn of the wrist. I think of the internal plumbing and wiring of his home, all lovingly re-engineered with his own hands. I think of the dumbwaiter he installed between the main floor and the "garden apartment." And I think of that same practical spirit expressed in the Othmer still, in his consultancies in Burma and Japan, in his patents, in his travels, and in his publications.

A sixth fundamental aspect of Don's personality was his pride in his chosen profession. To be an engineer—and above all a chemical engineer—was to respond to the needs of mankind in the most honorable and practical way. Don identified closely with ACS and AIChE, with ACHEMA and the *Encyclopedia of Chemical Technology*, and with New York's Chemists' Club, Société, and SCI. He reveled in the medals, honors, and offices that came his way—not because of the personal glory but, rather, for the recognition that he was the honorable representative of an honorable calling.

Don's sense of the goodness and the grandness of the chemical enterprise led to an immediate, strong interest in the Chemical Heritage Foundation. When we inaugurated the annual Othmer Luncheon in his honor, his modesty made him troubled. But then—in a characteristic personal focus on the profession itself—he took pride and pleasure in the way the Othmer Luncheon drew together the different societies for which he had served as president or director. The point was not the *name* of the Othmer Luncheon but that he, Don, through the Othmer Luncheon and the Othmer Library to which it pointed, was the means of reuniting parts of the chemical community sundered by the forces of specialization.

The seventh and final facet of Don's character that will endure in our memories is his generosity. His many students know the reality of his and Mid's hospitality and of his pains with their work. There is generosity—and the outlook of the academic engineer—in his words in *Who's Who in America*: "The reward to the educator lies in his pride in his students' accomplishments. The richness of that reward is the satisfaction in knowing that the frontiers of knowledge have been extended."

Don's generosity will live through his benefactions—to educational and religious organizations in Omaha, to Polytechnic University (what other major university has a member of its faculty as its largest single donor?), to the Brooklyn community (through hospital, church, and historical society), and above all through the Donald F. and Mildred Topp Othmer Library of Chemical History at Independence National Historical Park in Philadelphia.

Recording, studying, and making known the achievements of the chemical science: This is the high calling of the Othmer Library. The library is Don's most tangible legacy. Those of us who enjoyed the company and confidence of this consummate chemical engineer also treasure an intangible legacy. That legacy is the satisfying memory of Don's simple command and quiet humor, his combination of intellectual precision and control, his essential modesty, his probity, his practicality, and his pride in his profession. Don, we miss you.

Gerhard Frohlich

(Presented at Donald F. Othmer's Memorial Service, 4 December 1995)

Dear friends and all of you who mourn with us today in this historic church: I am honored to add my voice of remembrance to those distinguished speakers and friends of Don Othmer. If I may, I'd like to speak for the many students, and especially the foreign students, of Don Othmer.

There are probably only a few chemical engineering educators who had such a large number of students from around the world. These students, from India, Japan, China, Germany, and Israel, and many other countries, were attracted by Professor Othmer's reputation as a chemical engineering authority. This great reputation induced them to come to this country—I was one of them.

Don cared deeply for his students and in many cases took a very personal interest. From the time I was his teaching assistant, throughout my professional career, and until his death, Don was a mentor and friend to me with all that this implies. I am sure that my life would have taken quite a different course without Don's continuous interest and encouragement.

Don Othmer's impact on the chemical profession and the life of his students cannot be fully assessed by normal measurements, although they are readily recognizable in many ways. It has been said that the ultimate tribute to a man's life is not just to be found in formal eulogies or the written word, but in the unspoken, felt affection of people who were enriched and own a larger vision of the future because of him.

Don Othmer had that larger vision of the future. His inventive mind revolved around things that could be and then encouraged his students to reach for them. Biochemical engineering and unconventional renewable energy sources were subjects of his inventions long before they became popular subjects of study.

His was a meaningful life blessed with the love of a wonderful wife. Mid and Don were always wonderful hosts, and it was a privilege to be invited to their lovely home. We were touched by their love and devotion to each other. Over the more recent years, as Mid's illness became progressively worse, Don's care and patience were very moving.

My remembrance would not be complete if I did not mention the last few years when he struggled to keep out of the hospital and to stay in his beloved home on Columbia Heights. There were a number of friends who banded together to help him to make this possible, but there is one person who went beyond the call of duty. That was Bobbie Baxter, who cared for him and fought with him around the clock when his life was just hanging on a thread several times in the last two years. I believe it is in Don's spirit when I say thank you, Bobbie, for all you have done for him.

So, as we honor Don and remember his life, we must say that it was a long and beautiful life—although not absent of sorrow—which left lasting legacies to our profession and to all of us here.

In closing, I would like to read to you some excerpts from letters of condolences from some very prominent chemical engineers in China, which, I think, reflect the affection chemical engineers around the globe felt for Don Othmer. These condolences were transmitted to us through Dr. Teh Cheng Lo, who also was a student of Don's.

> Professor Donald F. Othmer was one of the most outstanding professors and the greatest leaders of our time in the area of chemical engineering. We are deeply saddened by the loss of such a great teacher and friend who . . . made great contributions to the development of coal chemical engineering in China. . . . His farsightedness is highly respected by my colleagues here and myself.
>
> —*Dr. Teng Teng*
> *Vice President, Chinese Academy of Social Sciences, Beijing*

> I and my colleagues are deeply saddened by the death of Dr. Donald F. Othmer. Dr. Othmer was a beloved teacher, a humanitarian, and certainly one of the world's leaders in chemical engineering. . . . His innovative achievements will always enlighten the progress of chemical engineering education and chemical industry.
>
> —*Professor Jia Ding Wang*
> *Member, Chinese Academy of Science, Tsinghua University, Beijing*

> For decades, Dr. Othmer has been the idol of Chinese chemical engineers. . . . We have lost a beloved teacher and one of the greatest leaders in chemical engineering. . . . His friendship towards China and Chinese people also gave me a very warm feeling.
>
> —*Professor Dongdi Wu*
> *East China University of Science and Technology, Shanghai*

Dr. Othmer's contributions to the field of chemical engineering has benefited the lives of many nations in the world. . . . He will always be remembered as a warm human being, a great teacher, and an innovative engineer. His legacy will live forever and we all will miss him.

—Patrick Wong
Senior Engineering Manager, Bristol-Myers Squibb Company

I think we all join Patrick Wong in saying—We will miss him.

George Bugliarello

(Presented at Donald F. Othmer's Memorial Service, 4 December 1995)

Rev. Blackburn, Mrs. Peterson and other members of the Othmer Family, Dear Friends: How can we today, in a few minutes, even begin to recall all that was great and unique about Don Othmer—about a man greater than life, a man who influenced the lives of thousands of students, a man who with his research, his writing, his inventions, and his consulting indelibly affected the engineering profession and the lives of virtually all of us? How can we?

The best we can do is to recall a few episodes of his rich and long life—a life that he shared so fully with Mid, who, unfortunately, is confined to her clinic in Omaha.

And as we recall those episodes of Don's life and mourn him, we can only say to ourselves that we are also celebrating the great and mysterious phenomenon of a God-given creativity that through Don Othmer and other exceptional individuals has so greatly enhanced our human reach and enriched our human experience.

Don Othmer was born ninety-one years ago in Omaha. He graduated from the University of Nebraska and earned a Ph.D. in chemical engineering at the University of Michigan. He worked for five years at Kodak and then joined the faculty of Polytechnic in 1933. In 1937 he became head of the Chemical Engineering Department—a position he held for twenty-four years, until 1961, when he was made the first distinguished professor in Polytechnic's history.

At Polytechnic, among his research achievements was the development of a method for correlating and predicting chemical engineering data—the Othmer Reference-Substance Method, which enormously simplified calculations. He also invented processes used all over the world by chemical and other industries and was granted some 130 patents. I know that some of the other speakers will talk further about this.

The growing reputation of Don Othmer was made even more towering by over 350 scientific papers and by the *Kirk-Othmer Encyclopedia of Chemical Technology,* which became the standard reference book in the field. I remember, when I first visited China in 1974, still during the Cultural Revolution, the excitement with which, at university after university, knowing that I was from Polytechnic, I was shown their prized possession, the *Kirk-Othmer Encyclopedia.*

Several of Don Othmer's students achieved their own great prominence, launched in their careers by Don's dedication to his calling. Only a month ago one of his former undergraduate students, Martin Perl, received the Nobel Prize in physics. As Don once said, "The reward to the educator lies in his pride in his students' accomplishments." This was indeed Don Othmer—and, as the *New York Times* said, his work as teacher, researcher, and department chairman helped transform Polytechnic into one of the nation's leading technological centers.

Recalling his first year at Polytechnic, Don Othmer once said, with the wry smile that we so well remember, "I came to work in September 1932 for $2,600 a year, and the Depression deepened that year. Spring of 1933 was a low point; even professors and heads of departments were leaving involuntarily. I was given a $100 a year salary cut, but to show it was not meant, I was raised in rank from instructor to assistant professor!" Already a leader, within a year of joining Polytechnic, he launched Poly's first graduate course in chemical engineering.

Don also took pride in his physical stamina. In 1987 he wrote me that in fifty-seven years at Poly he had had to stay home only twice for disability. The first was in 1954, when, in his words—and I see him smiling when he wrote this—"A great surgeon took my heart out, peeled off a quarter-inch casing, and put it back." It became legend at Polytechnic that a few days after the operation, and still in his hospital bed, he was working feverishly on a paper—the unsinkable Don Othmer!

In June 1977 Polytechnic conferred on Don an honorary degree, in recognition of his "prodigious and seminal contributions." In the eighteen years since, Don continued to write, teach, and create, with his mind active to the end. One wishes the laws of nature could have made an exception and given us even more of Don Othmer, but as we say our last farewell to him, we are thankful for all the priceless gifts that God gave us through Don.

James Kingsbury

Don was an inveterate storyteller. He used anecdotes every chance he had to make a point. At many hospital board meetings his witty comments and anecdotes changed desultory environments into positive ones. Don was constantly reminding us that we had to "look at the forest instead of the trees." That comment, often repeated at Finance and Joint Conference Committee meetings, would bring some wandering participants back to the important venue of trustee roles in maintaining the hospital's vision, mission, and level of quality. Don would not let petty disagreements among physicians last long. Patient care and quality delivery were his watchwords.

Don and Mid thoroughly enjoyed the elegant and low-key worship environment at Plymouth Church. They regularly sat in the middle of the sanctuary, near the front. Arnold Ostlund's organ playing was a joy for them. It was a custom for most of the congregation to stay seated at the end of the worship until the organ postlude was finished. Somehow, I think Don and Mid might have started that custom. They were always most intent as they sat and listened to Arnold's magnificent music.

Don's neighbors in Brooklyn Heights included many famous and well-known people. They were always friendly and saw Don as a quiet, unassuming college professor. If only some of those people had known his net worth at that time, they might have had a different image or assessment of Don. He had neighbors who would try to showcase their wealth in various ways—celebrity parties, stretch limousines, literary gatherings, elegant black-tie dinners, etc. Often times I would see Don on Columbia Heights the day after these events. He would comment to me about some of the extravagances and trimmings of wealth that were displayed the previous night. Such largesse was definitely not his style. He did not criticize any of these people; he just liked to reflect on the contrast of their lifestyles to his. Frankly, I think Don would have rather fixed an electric switch or cleaned the windows of his magnificent brownstone than participate in some of his neighbors' extravaganzas.

Don was usually modest about his investment prowess. Not many years after I met him, he asked me, since I was working on Wall Street, if I knew a "bright young boy" by the name of Warren Buffett. He said he got to know him when he was teaching in Omaha and that he was quite impressed with Warren. He never let on to the fact that his early investment with Warren was doing so well or that Warren was a close friend. Warren almost showed up at Don's eightieth birthday party, only one of Warren's less-than-successful investments, an investment in Salomon Brothers, prevented him from attending. But he wanted to be there to help Don celebrate.

Despite his closeness to Warren Buffett, his investment conversations with me often turned to utility stocks and whether or not some of them might cut their dividends. He often told me he was a lousy potential customer as he always reinvested his dividends and had no idea what his adjusted costs were. As such, he never wanted to sell anything as the capital gain's computation would not be worth the effort. His low-key approach toward his investments was again indicative of his general philosophy toward life.

Barbara J. Kohuth

I first met Don Othmer in 1982 when I joined the administrative staff of the Long Island College Hospital (LICH) and attended meetings of the Board of Regents. Don and I soon became fast friends, and this simpatico broadened to include my family when my husband—recalling his undergraduate days at Polytechnic Institute—remembered Don fondly and Don, happily, remembered him. Thereafter, we all enjoyed many family-style Sunday brunches together during which my father enjoyed keeping Mid's glass filled with mimosas (Mid loved to joke that she needed the orange juice to stay healthy) and talk of science and engineering dominated the conversation. Don and Mid later often referred to these comfortable, happy occasions. (Don was then retired but still walking to Poly each day to work in his office.)

Over the years both Don and Mid expressed their interest in helping me in my role as vice president for Development at LICH. After much discussion they revised their wills in the mid-1980s, and thereafter they both carefully edited a question-and-answer article, which we printed on the cover of our financial planning newsletter. In my interview with them they urged others to follow their lead in providing for the Long Island College Hospital. To help kick off a national health care fundraising campaign, "Give To Life," sponsored by the then National Association for Hospital Development, they agreed to let me announce an eight-figure commitment to LICH. When I suggested that the announcement would have more punch with a cash donation as well, they agreed to a $1 million cash gift. On another occasion, to help us raise money for a cancer center, they agreed to give another cash donation of $1 million as a challenge gift to encourage matching gifts. Don expected and received almost daily accounting of our progress and rejoiced as we succeeded.

As a member of the hospital's Financial Operations Committee, Don was concerned about the financial status of the hospital, and, as he told me many times, he foresaw problems ahead in health care and for LICH. His insights and positions were remarkable. His very forward-looking views included rights of women, women's role in leadership, the need for global communication, the right to die by request for the terminally ill, and a number of other social issues. I think Don could best be described as a fiscal conservative and a social liberal.

Mid was truly deferential to Don, but with her keen intellect and sharp wit, she had the cunning ability to get in the last word with a biting comment, said softly. Taking it good-naturedly, Don never missed a beat and both would then laugh. Mid was also most gracious, with a distinct sense of old-fashioned etiquette. She was a very private person, so I took it as a big compliment when she often insisted that I sleep at their house in Brooklyn Heights rather than take the Long Island Railroad back home after a very late meeting or when I needed to work late in my office. Also Mid's flower arrangements were among the most artful I have ever seen and graced the treasures they displayed from their trips to Japan and all over the world. Their home reflected their many travels and was almost like a private museum. Prominently displayed in the living room, of course, was the model of the Othmer still.

My most memorable conversations with Don took place in 1993–1994 during my long recuperation from an accident. Don phoned me at home every evening at 9:30 P.M. for eight months and chatted about the news of the day and anything else that came to his mind, including the high cost of living (he could be quite frugal and thought prices should be what they were fifty years ago!); current events; family and friends; travel memories; the politics at LICH, PolyTech, and Chemical Heritage Foundation; Warren Buffett; his stocks; and his overdue IRS return. His spirit was so rich and caring, and his thoughts so deep and penetrating. He was the most buoyant or depressed after having just spoken with Mid, who was in a nursing home by that time. It was during this period that I truly got to know the man and began to understand the depth of his complexity.

His intellect and financial acumen granted, he also had a very simple side and liked to putter about the house. I remember one day pleading with him when he was almost ninety years old not to climb up on the roof to fix a leak reported by a tenant. We had an argument about this over the phone, and in exasperation I accused him of not wanting to pay a repairman. He quickly countered that he could not trust anyone else to repair his roof. An hour later he phoned me triumphant in that he had accomplished the mission himself and the roof was repaired. Then he lowered his voice and sheepishly promised not to do it again because, "I noticed that the old legs are not as steady as they used to be." In the house at this time he was using a walker!

Of special note is the vision of Mid and Don Othmer in helping the institutions they chose to support. It constantly impressed me how steadfast they were to these institutions despite real upsets from time to time with the politics, the people, or policies of these organizations. They took the long view and did not allow momentary slights, however poorly masked and hurtful, to deter them. Herein was truly the essence of the philanthropic spirit. I think by far the most fun Don got out of his donations occurred when the Othmers gave Plymouth Church a healthy sum to refurbish the church organ. Don enjoyed this to no end and laughed heartily—the deepest laugh I ever heard him laugh—as he recounted his version of how a donation helped with "rejuvenation of the Othmer organ."

As the years progressed and Don's health faltered, I could see him begin to lose his tight, controlling grip on life. Don was aware of his own fragility. It was heartbreaking to watch him wrestle with the very difficult decision he had to make when he recognized his inability to care for Mid at home and realized that he would probably die before her. Lovingly, he took steps to ensure a safe and secure environment for Mid. But his resulting depression was palpable and his own health deteriorated. The late Robert B. Wolf and Ciril J. Godec provided at-home medical care while Sheldon H. Putterman served as a medical adviser. His home-care attendant, Bobbie Baxter, was most devoted, and his attorney, Ted Wagner, was always ready to assist. As Arnold Thackray and I kept daily tabs on Don, the Long Island College Hospital stood ready for medical backup. None of us was ready to let Don go. But the end came, sadly.

These are just a few of my recollections of Don and Mid Othmer. Knowing them was like viewing a complicated tapestry with meaning unfolding strand by strand—and with a resounding impact, especially when one stood back to view from afar. My life has been truly enriched by knowing Mid and Don, and the institutions they supported will be forever changed. I only hope that these institutions handle the Othmer largess with the fiduciary care befitting one of the most philanthropic couples in the history of our nation.

Tranda Schultz Fischelis

When I think of my aunt and uncle, Mid and Don Othmer, I think of the many Christmas days we spent together. When I was a child, Mid and Don would come to visit my family in Lincoln, Nebraska. Don was my mother's older brother.

One Christmas I received an unassembled metal dollhouse as a gift, and it took two Ph.D.s to put it together. My uncle and my father, who was a professor of paleontology at the University of Nebraska, sat on the living room floor and read the directions before meeting their challenge. After struggling for a while, both agreed to ignore the directions and to use common sense. Although Don was not completely satisfied with the final product, the house stood for many years.

Even as a child I could see that under that smile and quick wit, Don was a serious man, serious about making our world a better place in which to live. During a typical Christmas dinner conversation Don would comment about a new Othmer-developed technology or processing method, such as the desalination of water and the liquefaction of coal. At the same dinner table there was an ever-present balance to the practical world of chemical engineering. My father, the geologist and paleontologist, would tell of his latest fossil discovery and how it fit into the evolution, migration, and extinction of fossil mammals in the Great Plains. One year the conversation settled on a discussion about how to build a concrete canoe. I believe one of my mother's famous desserts interrupted and ultimately ended this would-be invention.

Years later, when I was in advertising and then a product manager at Rohm and Haas, colleagues asked me how I stood up so well during the long, technical meetings. I would respond, "They sound just like conversations with my uncle and father over Christmas dinner." No one would believe me!

After college I moved to Philadelphia, a two-hour drive from Brooklyn Heights, where Mid and Don lived. For years my husband, daughter, and I were invited to attend Christmas dinner at the lovely Victorian brownstone, which overlooked the East River and Manhattan. The roles for the day were well defined. Mid, a fabulous cook, was the chef; Don carved the turkey; and Mid's older sister, Alice Topp-Lee, artistically placed the food on the plates.

The guest list was always of interest. Regulars included Jack Shad, former chairman of the Securities and Exchange Commission and ambassador to the Netherlands during the Reagan administration, and his family. Another guest was Alice Topp-Lee's friend, David Ford, who starred in the Broadway musical *1776*. And then there was Alice, a fashion designer, who created the junior miss line of clothing after World War II.

When Mid's and Don's health began to fail, they were forced to give up entertaining at Christmas. When I was in New York on business, I would stop by to see them, or I would take a train ride on Saturday to New York and a cab to Brooklyn to make the visit. On one of these trips I remember Mid and Don talking with great enthusiasm about the new headquarters for the Chemical Heritage Foundation. When the subject changed to teaching, Don asked Mid what it was like to teach high school English in Nebraska. Mid responded, "Life did not begin, Don, until I married you."

Before marrying Don, Mid was a former high school teacher who received her master's degree from Columbia University, and she was a buyer for her mother's dress shops in Omaha. When they married in 1950, Mid and Don agreed that if he were ever to be away from home for more than three nights, she would go along. I remember in the 1970s Mid and Don telling me that they had been around the world thirteen times.

In October 1995 I called Don to check in with him. (By then Mid was in a nursing home.) He immediately asked, "Are you coming up to see me?" And I responded, "Yes, I am." And he said, "Well, you had better hurry." I took the train up, and I am glad I did. That was the last time I saw Don. One week later he passed away.

Mary D. Seina

My Aunt Mildred was born on 13 September 1907—a Friday. She always told people what foolishness the superstition about Friday the thirteenth was. She was the second of three daughters of Holgar and Mattie Topp. Her sister Alice was five years older and my darling mother, Nelsie, five years younger than she. Shortly after my mother's birth Mattie and Holgar were divorced, and my grandmother, a former schoolteacher with a hearing loss, became a single parent in 1912. To support her family, Mattie bought a yard goods store in Omaha on 17th and Vinton streets with an apartment on the second floor.

The girls grew up in an almost Louisa May Alcott tradition: All shared in housekeeping responsibilities and tended the store under the tutelage of a very nurturing and loving mother.

My Aunt Alice did the sewing and helped with the dressmaking—a career she followed through life. She left home at age nineteen and became a very successful designer of clothing in the United States and in Europe.

My mother watched the store, did the shopping, and tended to her mother—a pattern that would follow through her life. She continued to operate and expand the Topps' stores from yard goods and bib overalls to the new ready-to-wear and through many locations and successes. My grandmother lived with our family all her life.

My Aunt Mildred managed a lot of the kitchen duties. She was a wonderful cook. She was also a voracious reader, and she was always tutoring and assisting girls in their academic endeavors. Academics were clearly my aunt's forte. She graduated from the University of Nebraska and received her master's from Columbia. She taught English at Benson High School, where she developed a deep friendship with fellow teacher Alice Buffett. Teaching, especially English, always came naturally to my Aunt Middy. She was very helpful to me in my childhood and during her last five years in Omaha. Aunt Mid also helped her devoted companion, Naree Pankey, a woman born in Thailand who had obtained only a fifth-grade education. Naree's culinary skills flourished during her time with Aunt Mid.

As the Topp store continued to grow, my grandmother, my mother, and my Aunt Middy began taking buying trips to New York. Soon my aunt decided to become a resident buyer, and she joined forces with the Mildred McGohon buying office. She leased an apartment in Tudor City overlooking the East River.

Mid continued to come home to Omaha for holidays and my grandmother's birthdays. The traditional celebrations with my parents continued throughout their lives. Christmas always meant Alice and Middy coming home. In their later years my aunt and uncle spent the holidays at my parents' home in Rancho Mirage, California.

I remember Mid bringing Don to our annual family Christmas party in 1949 and all the excitement that caused among the sisters. Mid and Don were married the following Thanksgiving in New York. It was one of the highlights of my childhood—not only was I honored to be the flower girl, but I also saw the famous Macy's Thanksgiving Day Parade.

My aunt moved into my uncle's brownstone at 140 Columbia Heights where they enjoyed a wonderful life. My aunt kept a beautiful home and garden, cooked, entertained, and delighted Don's friends, students, and business associates. There was a parade of guests through the Othmers' drawing room.

Their lives were filled with much happiness and mutual respect. They shared many interests, including a great love for their home, travel, and academia. They wrote a fabulous travelogue of their many journeys—the "Dear Folks" letters—and ended the letters with their signature closing: "Cheers, Midon." Hundreds of people received the Dear Folks letters and, at Christmas, the wonderful plaques that they designed and had made in Japan depicting some part of their lives.

I have wonderful memories of my aunt and uncle. They both taught me a great deal. My aunt nurtured my love of books, miniatures, cooking, and pets. Of her many lessons to me on grammar, etiquette, and life, two come instantly to mind:

1. Like little ships that sail to sea, I steer my spoon away from me.
2. It takes a very good husband to be better than none.

You can tell we covered a wide spectrum.

We always spent a lot of time reminiscing. Aunt Mid deeply loved her family and was able to meet both her beautiful grandnieces and to enjoy all of her family during her final years. She was a brilliant and beautiful woman. The last five years of her life were an opportunity for me to give back to her—one for which I will be forever grateful.

Leslie Shad

(Presented at Mildred Topp Othmer's Memorial Service, 17 September 1998)

Mid was my godmother. She became my godmother after she met my parents, as their landlord. When my parents moved to New York City, Mid and Don rented the top floor apartment of their townhouse to my parents. Eventually my parents were spending weekends at Mid and Don's Pennsylvania farm and ice skating with them on the pond near their house.

From the time I was very small, Mid visited our country house in Massachusetts. Mid, my mother, grandmother, and I would picnic beside a fast, cold stream. Afterward, Mid would collect her skirt in her hands and go wading with me in the quick current. We would always get splashed. Sometimes we would fall down.

In a lake beside our country house, Mid would swim and water ski with us. Along the lakeshore, she would collect blackberries with me, sharing the thorns during the picking and the berries afterward.

During the holidays we would have dinner with Mid and Don at their Brooklyn Heights home. Mid would load the soup on the dumbwaiter and let my little brother or me pull the ropes to send it downstairs for appetizers in their sitting room. Sometimes when we were very little, Don would give my brother and me a ride on the dumbwaiter, but it made Mid a little nervous.

When the grownup talk at holidays became too long and chemical for my little brother and me, Mid would pull out a stash of toys. There was a monkey with clanging cymbals that we called Jocko, a white barking dog, a sock chimpanzee, and more. Mid would discreetly play with us, continuing to talk with the adults but also pointing out hiding places for the toys. She showed us that the latticework of a wood table made a perfect cave for our toys. There was a round metal table that made a resounding bang when a toy or child's fist landed hard. Then if it was not too cold outside, Mid would take us on a tour of the sculptures and Japanese lamps in her garden. If it was too cold, we would visit her menagerie of netsuke and other carvings and turn on and off the light in the translucent sculpture beside their front doorway.

In December their porcelain Christmas card with a seasonal greeting from "Midon" would arrive, and I would receive a present of a scarf, a handkerchief, or a piece of lace.

In later years I would spend many Saturday mornings near the Othmers' townhouse escorting Planned Parenthood patients through lines of protesters into the Brooklyn Heights clinic. Afterward, I would stop by to visit the Othmers. Mid and I were mutually interested in Planned Parenthood. But most of the time on these and other visits we would talk about her memories and her cat. She had taken in a small, very furry brown kitten. She loved the cat dearly, even after Don noticed that the cat shed profusely and had forsworn the litter box for the furniture. Don exiled the cat, but Mid continued to pine for it.

Mid was a thoughtful and very loving person. I cherish my memories of her, especially her gracious gestures and her kindnesses toward children and small animals.

George Bugliarello

(Presented at Mildred Topp Othmer's Memorial Service, 17 September 1998)

Mildred Topp Othmer—or Mid, as she was affectionately called—was a gentle soul who, throughout her married life to Don, gave him emotional support and companionship. This emotional support and companionship tempered Don's almost superhuman drive to excellence and commitment to work.

I came to Polytechnic in October 1973. Naturally I had known of Don much before that and was awed by his reputation. When Mid and Don invited my wife and me to their house, we had the pleasure of meeting Mid. We saw how well, indeed, she complemented Don, and Don her. Don, of course, tended to guide the conversation—and it was, as always, an interesting and challenging conversation—but Mid joined in with soft-spoken recollections of the trips they took and enjoyed so frequently. She was also, in her quiet way, very solicitous of us.

After that, over more than two decades, Virginia and I were many times guests at dinners of understated Mid elegance, and we had the pleasure of having them at our house. What I remember so vividly of Mid at these and other occasions was her pride in Don and her support of him and, it must be said, her patience when, at public events, he gave very long talks, as he was often wont to do.

With Don, Mid's strong support of Polytechnic, which had been Don's academic home through most of his career, was generously manifested in her will and helped build excellence at the university. But Mid's devotion to Don did not diminish the quiet intensity of her own commitment to things that were exclusively hers—like Planned Parenthood and the Brooklyn Botanic Garden. These interests did not exclude other interests she had in common with Don, like the Othmer Library, the hospital, and Polytechnic; rather, they complemented those interests. And with Don she took great pride in recounting the story of the many interesting books and art pieces in their house.

Her concerns with Planned Parenthood became undoubtedly sharpened by what she saw in her trips with Don to developing countries, particularly in the immediate postwar period. She used to describe the poverty and the exploding birth rate with a sense of dismay and yet with the hope that the situation could be improved—a hope accompanied by her personal engagement here in New York.

Mid was a lady of taste, which was so manifest in her gracious hospitality. Her taste was innate, but undoubtedly it was sharpened by working with her mother at the famous Topp stores in Lincoln, Nebraska.

And Mid was a lady of compassion and understanding, qualities that must have made her an excellent teacher. Her background and her bent for teaching gave her a beautiful diction and a felicity of expression that added to the attraction of the Othmers' hospitality.

Mid's graciousness remained even when disease began to incapacitate her in the latter part of her life. On my visits to the Othmers in that period I was always moved by the fact that Mid, even if her mind could not always recollect or focus sharply, *never* abandoned her role as hostess. In sum, I believe it truly could be said of Mid—and Don—as Shakespeare said:

> He is the half part of a blessed man,
> Left to be finished by such as she;
> And she a fair divided excellence,
> Whose fulness of perfection lies in him.
>
> —*King John,* Act 2, Scene 1

Warren E. Buffett

BERKSHIRE HATHAWAY INC.
1440 KIEWIT PLAZA
OMAHA, NEBRASKA 68131
TELEPHONE (402) 346-1400
FAX (402) 346-0476

WARREN E. BUFFETT, CHAIRMAN

September 8, 1998

Mr. Theodore R. Wagner
Carter, Ledyard & Milburn
2 Wall Street
New York, NY 10005

Dear Ted:

Thanks for the invitation to the memorial service. I have to be in Omaha on that day, but my thoughts will be with Mid.

I came to know Mid through two different sources; I can't remember which was first. Her mother, Mattie Topp, was a marvelous woman, and I first called on her to sell her securities in 1951 when I was 21. At some point she introduced me to her daughters. However, my Aunt Alice — who was an enormous favorite of mine — taught with Mid at Benson High in Omaha and told me about her in my earlier years. At 68, I can't remember whether Alice or Mattie was responsible for the initial introduction.

Mid and Don were the perfect partners — supportive, never second-guessing and patient. They deserved every penny that evolved from their original investment.

Sincerely,

Warren E. Buffett

WEB/db

Their Legacies

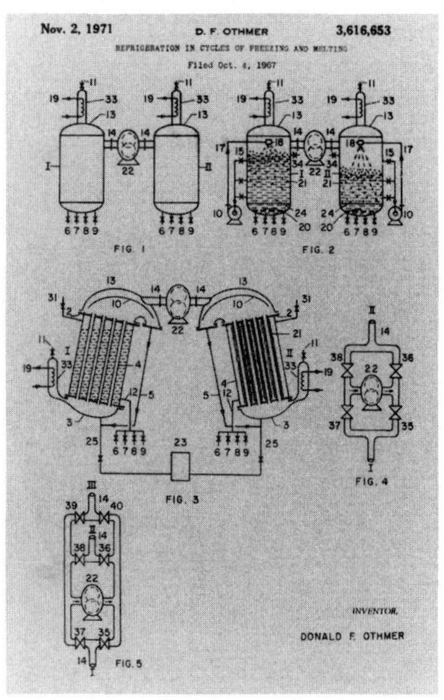

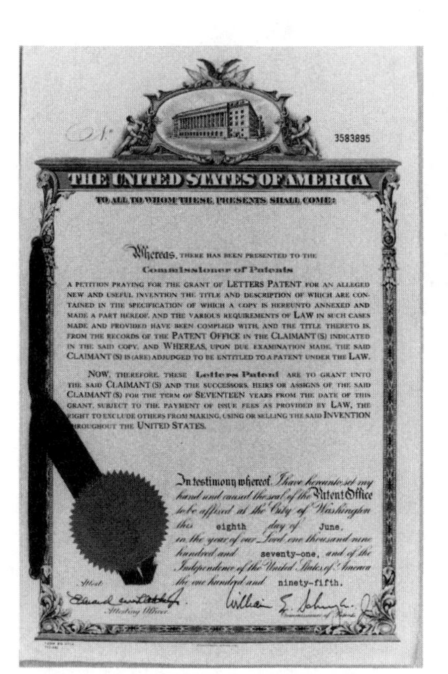
No 3583895

THE UNITED STATES OF AMERICA

TO ALL TO WHOM THESE PRESENTS SHALL COME:

Whereas, THERE HAS BEEN PRESENTED TO THE Commissioner of Patents A PETITION PRAYING FOR THE GRANT OF LETTERS PATENT FOR AN ALLEGED NEW AND USEFUL INVENTION THE TITLE AND DESCRIPTION OF WHICH ARE CONTAINED IN THE SPECIFICATION OF WHICH A COPY IS HEREUNTO ANNEXED AND MADE A PART HEREOF, AND THE VARIOUS REQUIREMENTS OF LAW IN SUCH CASES MADE AND PROVIDED HAVE BEEN COMPLIED WITH, AND THE TITLE THERETO IS, FROM THE RECORDS OF THE PATENT OFFICE IN THE CLAIMANT (S) INDICATED IN THE SAID COPY, AND WHEREAS, UPON DUE EXAMINATION MADE, THE SAID CLAIMANT (S) IS (ARE) ADJUDGED TO BE ENTITLED TO A PATENT UNDER THE LAW.

NOW, THEREFORE, THESE Letters Patent ARE TO GRANT UNTO THE SAID CLAIMANT (S) AND THE SUCCESSORS, HEIRS OR ASSIGNS OF THE SAID CLAIMANT (S) FOR THE TERM OF SEVENTEEN YEARS FROM THE DATE OF THIS GRANT, SUBJECT TO THE PAYMENT OF ISSUE FEES AS PROVIDED BY LAW, THE RIGHT TO EXCLUDE OTHERS FROM MAKING, USING OR SELLING THE SAID INVENTION THROUGHOUT THE UNITED STATES.

In testimony whereof, I have hereunto set my hand and caused the seal of the Patent Office to be affixed at the City of Washington this eighth day of June, in the year of our Lord one thousand nine hundred and seventy-one, and of the Independence of the United States of America the one hundred and ninety-fifth.

Attest:

Attesting Officer

Commissioner of Patents

Left: Cover page. United States, Patent 3,583,895. Evaporation using vapor-reheat and multi-effects, 8 June 1971. *Above:* Detail. United States, Patent 3,616,653. Refrigeration in cycles of freezing and melting, 2 November 1971.

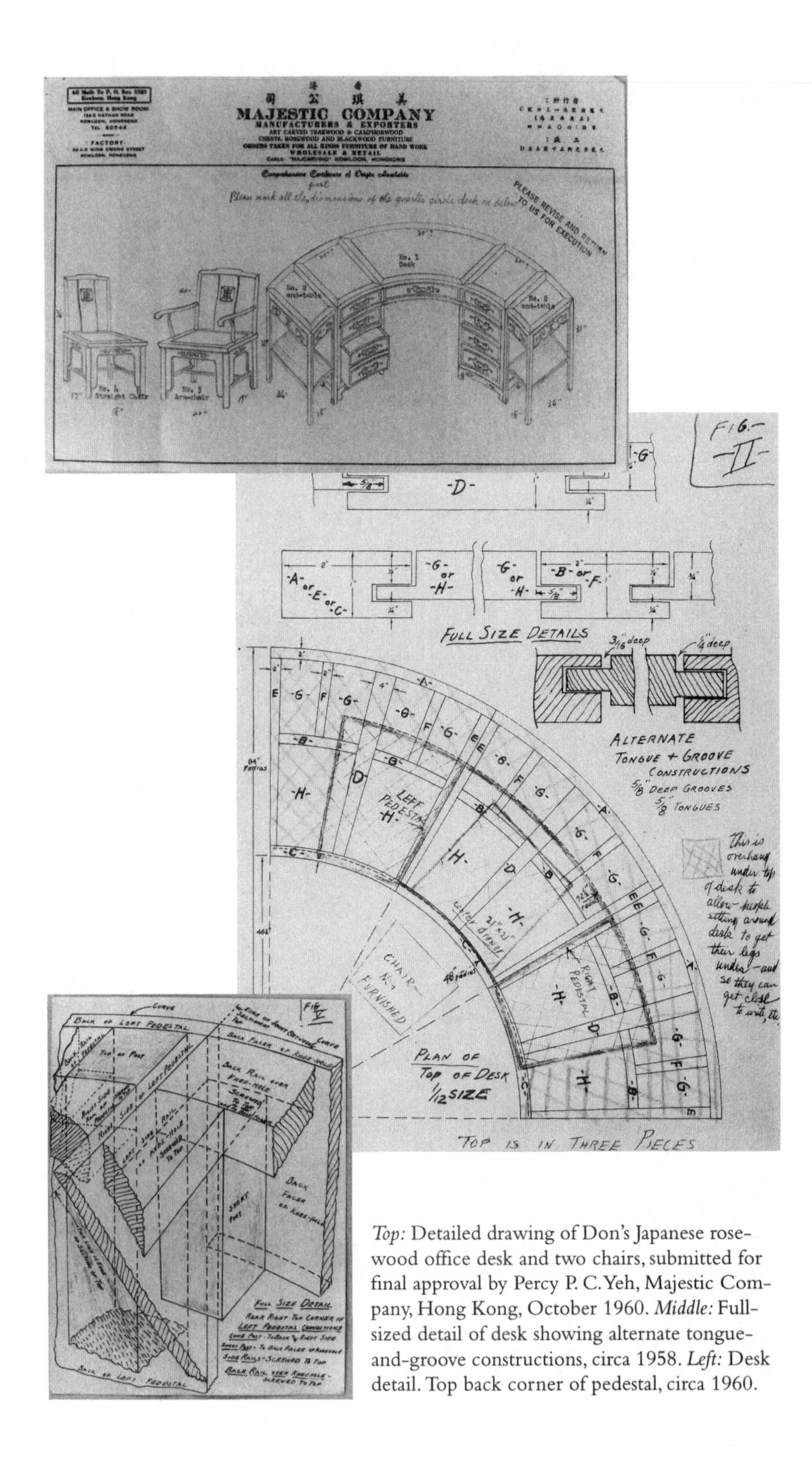

Top: Detailed drawing of Don's Japanese rosewood office desk and two chairs, submitted for final approval by Percy P. C. Yeh, Majestic Company, Hong Kong, October 1960. *Middle:* Full-sized detail of desk showing alternate tongue-and-groove constructions, circa 1958. *Left:* Desk detail. Top back corner of pedestal, circa 1960.

Section 4

The Othmer Archives at the Chemical Heritage Foundation

by Susan Hamson

"If there is one word I would use to describe Don," said Arnold Thackray, president of the Chemical Heritage Foundation, "it would be *practical*." Indeed, it would be hard to argue that the professional life of a chemical engineer would be anything but practical. Don's contributions to the field in desalination, distillation, refrigeration, petrochemicals, and waste management (to name but a few) were all profoundly practical and infinitely beneficial. But for Don, practicality seemed to be an inherent quality. Even a cursory glance at Don's record keeping and personal artifacts reveals this one certainty about him.

Among the most fascinating artifacts in the archives are four small, black, innocuous books found neatly stacked in an old shoebox; they are Don's personal diaries from 1921 to 1924. These diaries offer a peek into the life of a young man in Omaha who attended church every Sunday, helped his parents around the house, flirted with the girls, went canoeing with his friends, and loved chemistry. Tuesday, 22 February 1921: "Made the pendulums and pivots for a Harmonograph, another kind of designing machine. Tried to replicate zinc oxide with coke but it wouldn't work." In May of the following year Don lamented the mishaps that often come with lab work: "Went up to chem lab this afternoon, cleaned up, and checked out. My excess breakage was $4.20, which means $14.20, almost broke my pocketbook."

"I was an avid reader," Don remarked on his youth. "I did enjoy reading various books on the sciences and whatever came to hand otherwise." Don made a list of the books he'd read in 1921: *Modern Chemistry and Its Wonders*, *Romance of Scientific Discovery*, *The New Knowledge*, and *Chemistry in the Service of Man*. In high school Don noted that he had "been reading a book on the history of mathematics." His opinion of the work? "It is pretty deep."

Combining his love of history and books, Don collected some fifty old and rare texts in the field of chemistry and the related sciences. Among the collection that is now part of the Othmer Library is Hieronymus Brunschwig's *Liber de Arte Distillandi de Compositis* (Strassburg, 1512), *The Art of Distillation* by John French (London, circa 1656), and Vannuccio Biringuccio's posthumously published *Pirotechnia* (Venice, 1550). These books combine the skills of printing, illustration, and binding with significant content and rarity.

Don's interest in amateur photography is also revealed in his diaries. "Went down to one of the shops on 35th Street and bought that Kodak . . . I have been wanting for so long," he wrote at age nineteen. "It is a little beauty." A small box from his office at Polytechnic contained numerous photographs taken in the early 1920s. Few of the individuals in the pictures are identified, but Don can be seen in many of them canoeing, posing with friends, and just generally having a good time. Some of the pictures actually look like photographic experiments, reflecting Don's almost inherent desire to understand how things work. Silhouettes and subjects positioning objects and posing in a variety of lighting effects all characterize this collection. A small envelope specifically labeled "Girls" is certainly among the most interesting. Again, most of the subjects are not specifically identified, but one photograph is particularly amusing.

Sitting on the rear bumper of a Model A Ford is a lovely young woman who has signed the back of her photograph "with my regards"; she is Marie Price. In the corner of the photograph Don has written his impressions of Miss Price.

> Nice girl—good dancer—fair line—only fault she scatters attentions when with one man. Would a second date cure her?

Don may have been unique in his abilities as a chemist, engineer, and inventor. However, when it came to young women and youth, Don seems charmingly typical.

While personal artifacts and glimpses of his younger days are certainly part of the collection, it is Don's career at Polytechnic University and as an internationally renowned independent consulting chemical engineer—a career spanning almost seven decades—that makes up the bulk of the Othmer archives. Don began his academic career in 1932, but, as he once

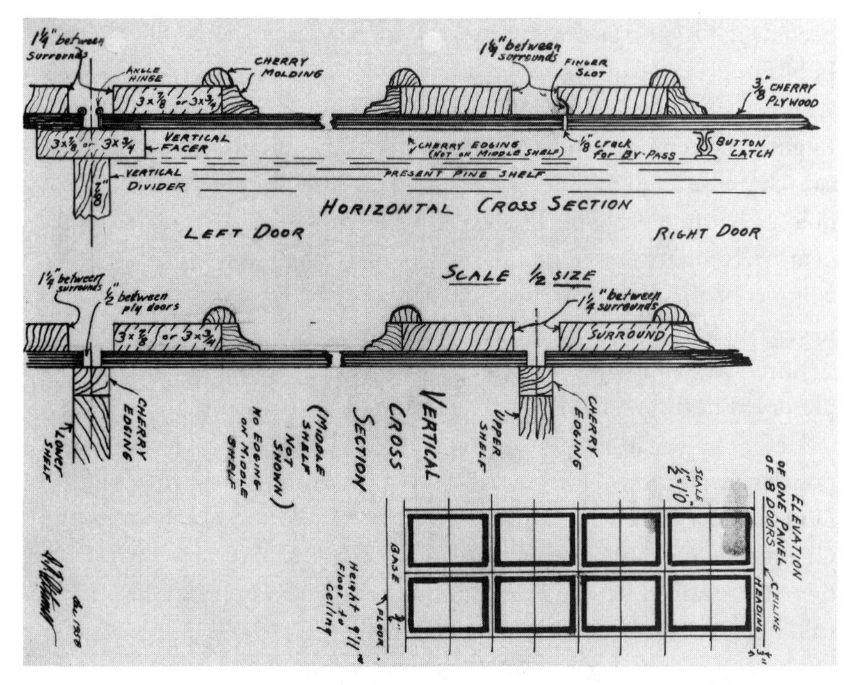

Detail for the shoji screens that covered Don's office windows at Poly, December 1958.

noted, "[I] took the job . . . in stride, although I never had the very real joy in this that many professors have expressed." Don's passion was in finding practical solutions to complex problems; through this search he formed lasting relationships with his students. "Students in general had a feeling of accomplishment . . . and I enjoyed working with them. After the final lecture, in two different years, at the final 'Good-bye' they gave me a hearty hand-clapping and cheers, the most heartwarming occasions in my many years of teaching."

Many of Don's student associations and friendships are represented in the archives, as are exams, dissertation topics, and some administrative information. Don directed the master's theses and doctoral dissertations of hundreds of candidates. Some of his notable students include Gerhard Frohlich, retired vice president of Hoffmann–LaRoche, Inc., and 1999 president of the American Institute of Chemical Engineers; Raphael Katzen, head of Raphael Katzen Associates International, Inc.; and Martin Perl, professor of physics at Stanford University and co-recipient of the 1995 Nobel Prize in physics.

In 1945 Don and Raymond Eller Kirk, then head of the Department of Chemistry at Polytechnic University, began collaborating on what would become the *Kirk-Othmer Encyclopedia of Chemical Technology* with Interscience Publishing Company (now an imprint of John Wiley and Sons, Inc.). Citing a need for an "encyclopedia of chemical technology, written in English and representing modern American practice and modern American chemical engineering," Don and Kirk produced a publication that quickly became the bible of chemical engineers everywhere. "In the Argentine," remembered Don, "the library only had . . . the first and second volume of the first edition of the *Encyclopedia*. These were so worn. I never saw books in a library that were so worn."

Some advertising literature from the first and second editions has been preserved in the archives.

Before he began his career in academe at Polytechnic, Don accepted a position as a junior engineer for Kodak in 1927. In the years that followed, Don played a leading role in the production of cellulose acetate at Eastman Chemical Company's plant in Kingsport, Tennessee. To observe how acetic acid is distilled, he designed what has come to be known as the Othmer still. For the first time chemists could accurately measure the difference between the concentration of substances in boiling liquid and its vapor. Don also furthered the science of azeotropic distillation, a process that introduces a third chemical during the distillation process to improve purity and reduce energy consumption. A mounted model of the Othmer still is part of the archival collection. This model is similar to the one shown in the oil portrait of Don that hangs in the Othmer Library's Reading Room alongside that of Mildred posed in her elegant blue wedding dress.

In the early 1930s, frustrated that the monetary compensation he received was low—especially in view of that fact that Kodak held more than 40 patents based on his research—Don left industry and entered the world of personal consulting. In the next sixty years Don acquired more than 150 U.S. patents, as well as scores of patents around the world. The archive is privileged to hold many of the original patents, as well as applications and accompanying documentation. The Japanese patents, in particular, are very striking, as they seem more works of art than legal license.

Don was a champion of large-scale desalination of ocean water and received a patent in 1975 that used sunlight and seawater to produce fresh water, electric power, and nourishment for clams, oysters, and other ocean life. Don also believed in using methanol as a cleaner alternative to gasoline. Documentation for these, as well as other challenging technical problems—waste management, heating arctic pipelines, increasing citrus yields,

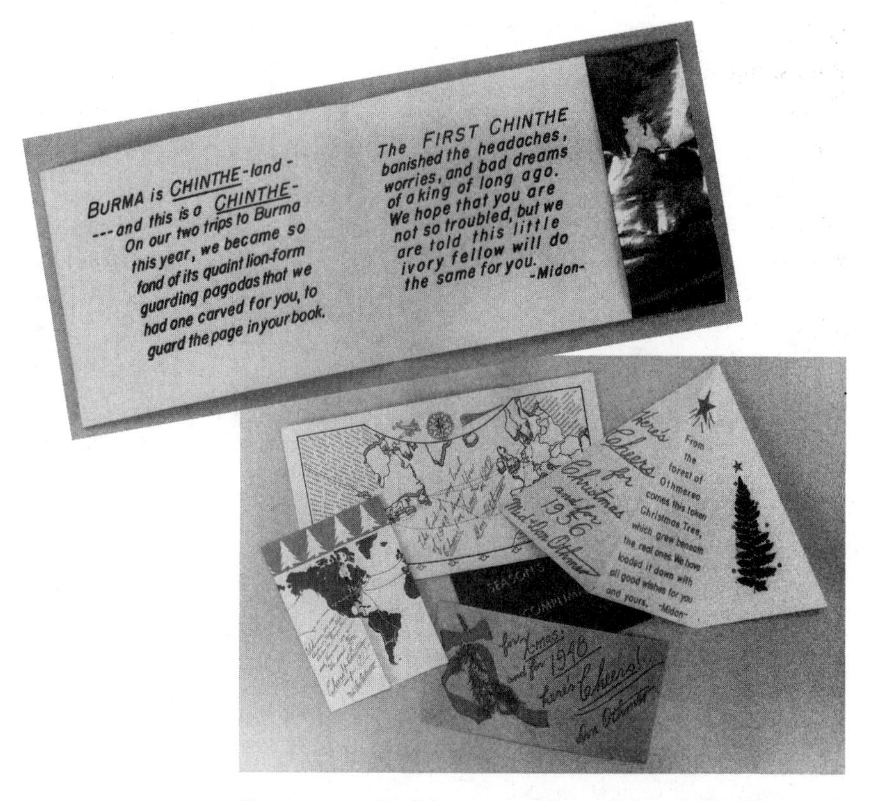

Top: A personal holiday greeting from Midon, 1950s. The carved two-centimeter high chinthe on the right is clipped to the card and nestled in a green foil ribbon. *Bottom:* A selection of personalized Christmas cards through the years. Even before Don and Mid were married in 1950, Don's propensity for holiday creativity was obvious. Clockwise from left to right: 1953, 1949, 1943, 1956, and 1948.

and protecting fuel tanks from corrosion—makes up a substantial portion of the Othmer archives.

The many companies for which Don consulted in Asia, the Americas, and Europe are all part of the voluminous correspondence that is preserved in the archives. With his renowned expertise as a chemical engineer, Don was in demand throughout the world. He was a regular visitor to Asia after World War II and, as consultant to the giant international engineering firm TAMS, helped rebuild the Burmese chemical industry in the early 1950s. Although he was asked to live in Rangoon for at least three years, Don declined and supervised most of the work from New York with assistants in Burma. "It was fascinating work," said Don of his

Two of Midon's famous ceramic holiday cards. The first, produced in 1960, is on the right. On the left, the impressive New York City skyline makes a dramatic backdrop for two Japanese cranes in 1968.

largest industrial project. And such groundbreaking expertise in private industry had not gone unnoticed by the U.S. government.

Earlier, during World War II, Don had been trained as an evaluator of the German chemical industry. In June 1943 Don received a telegram ordering him to appear for training. Along with fifteen other men Don was ultimately subjected to what he called "the third degree."

> We had an overhanging worry knowing that each was to be subjected to the third degree sooner or later. That ultimately did happen. . . . The interrogators were behind the spotlight, asking you all sorts of questions. . . . It was a wonderful game [but] they did it in a way as to be intentionally nerve-racking.

The purpose of this training was to prepare a select group of chemical engineers to evaluate German chemical military production in occupied plants in France and Belgium as the Allied forces invaded. However, the end of the war in Europe made this program unnecessary. Don continued to work as a consultant to the Chemical Corps of the U.S. Army through the mid-1950s. In 1951 Don was cleared by the Ordnance Corps of the army to work on the Scientific Advisory Committee and other consultation efforts with the Picatinny Arsenal. Of his contribution to the war effort, Don reminisced, "In one of my $1 per year jobs, I did get a check for 59¢; payment for 59/100 of a year's work. That was '44 . . . and I have the original check." Thanks to Don's meticulous record keeping, this

Midon's 1970–71 ceramic holiday greeting. " 'Flipper' is our bronze fountain from Japan, where a carp is the symbol of strength and courage—he swims against the current. In our garden overlooking Manhattan he leaps from a sea of ivy with an outpouring of our *Cheers!* for your *Christmas* and for *1971. Mid* and *Don Othmer*."

check—and two others from 1943 and 1945—are part of the Othmer archives.*

Don carried out several missions in Central and South America for the State Department, but there was plenty of work for Don on the home front as well.

> There were many domestic activities always requiring much travel. Not always, even with important war work could one get priority for air travel. . . . Pullman reservations were also tight—and I remember six or maybe it was seven *consecutive nights* in Pullman berths. Most of these were uppers, which were 73½ inches long, just my length. But I always slept on my stomach, which left my toes out in the aisle but behind the curtain.

Travel was such a necessary part of Don's professional life it is fortunate that it was also a welcome addition to his and Mid's lives. In 1947 Don began detailing his travels abroad in a letter to his mother; by 1949 the letter was a firmly established tradition—the "Dear Folks" travelogues addressed to family and friends. The tradition continued after Don and Mid were married in 1950. While business meetings, lectures, and conferences usually occasioned their travels abroad, the Othmers turned these

* These checks for fifty-six cents, one dollar, and fifty-six cents were issued in 1943, 1944, and 1945, respectively. The check to which Don refers was actually one dollar for that year.

opportunities into personal vacations. In a letter that is typical of Dear Folks, Mid and Don paint such a detailed portrait that the armchair traveler is taken along for the adventure.

> A magnificent eight-panel screen with a fine painting on gold was unfolded at one end of the room; and the number four daughter of the house, in a blue and white kimono and a red obi, performed with great concentration and grace a classical dance to the sad samisen music played on a phonograph. . . . That night we slept on the floor under a great square, pale-blue mosquito net. A violent storm with lashing rain and much lightning silhouetted a jagged mountain rising sharply just across the road.

The Othmers' admiration for Japanese life and culture is expressed many times in their letters, but it is also evident throughout the collection, particularly in several pieces of furniture from Don's office at Brooklyn Polytechnic and his study at home. Notable among these items are three desks: two identical desks where Don and Mid sat to work in their study at home and a large quadrant desk from Don's Polytechnic office.

The latter was unusual in many respects, not the least of which being that the desk is shaped as a quarter circle, with a pedestal at each end. Don not only drew the specifications for the desk's blueprint but also personally selected the Japanese rosewood from which it was ultimately handcrafted by the Majestic Company in Hong Kong.

> Since your drawings are so wonderful and mechanical that we believe we will manufacture the Desk with our whole heart and soul so that it will come out as you want it to be. . . . As regards the price for making the Quarter-circular Desk, we are very sorry to confess that we have never manufactured such a desk before so we do not have ideas about how long it will take our labourers to finish it.

The final product, received by Don in September 1961, was spectacular. With the custom-made shoji screens that covered his office windows, the desk made Don's office into a small slice of Japan in Brooklyn. Friends and colleagues who visited his Polytechnic office recall that Don placed a telephone at each end of his desk. As he left his office, he would use one phone to call the other, thereby generating a busy signal to callers on both lines. In the days before voice mail this arrangement ensured that Don never missed an important call.

Also illustrative of Don and Mid's love for the skill of Japanese artists and craftspersons are the couple's ceramic Christmas cards. However, these ceramic greetings were not without precedent. Found in a red and gold brocade scrapbook of Japanese design are a number of personally designed

Right: "Doc" Othmer lectures his students. "Bad data. Make it straight." *Below:* "My still sure makes good w[h]iskey." Don is fondly parodied by an admiring student. Both were presented to Don on the occasion of his seventieth birthday in 1974.

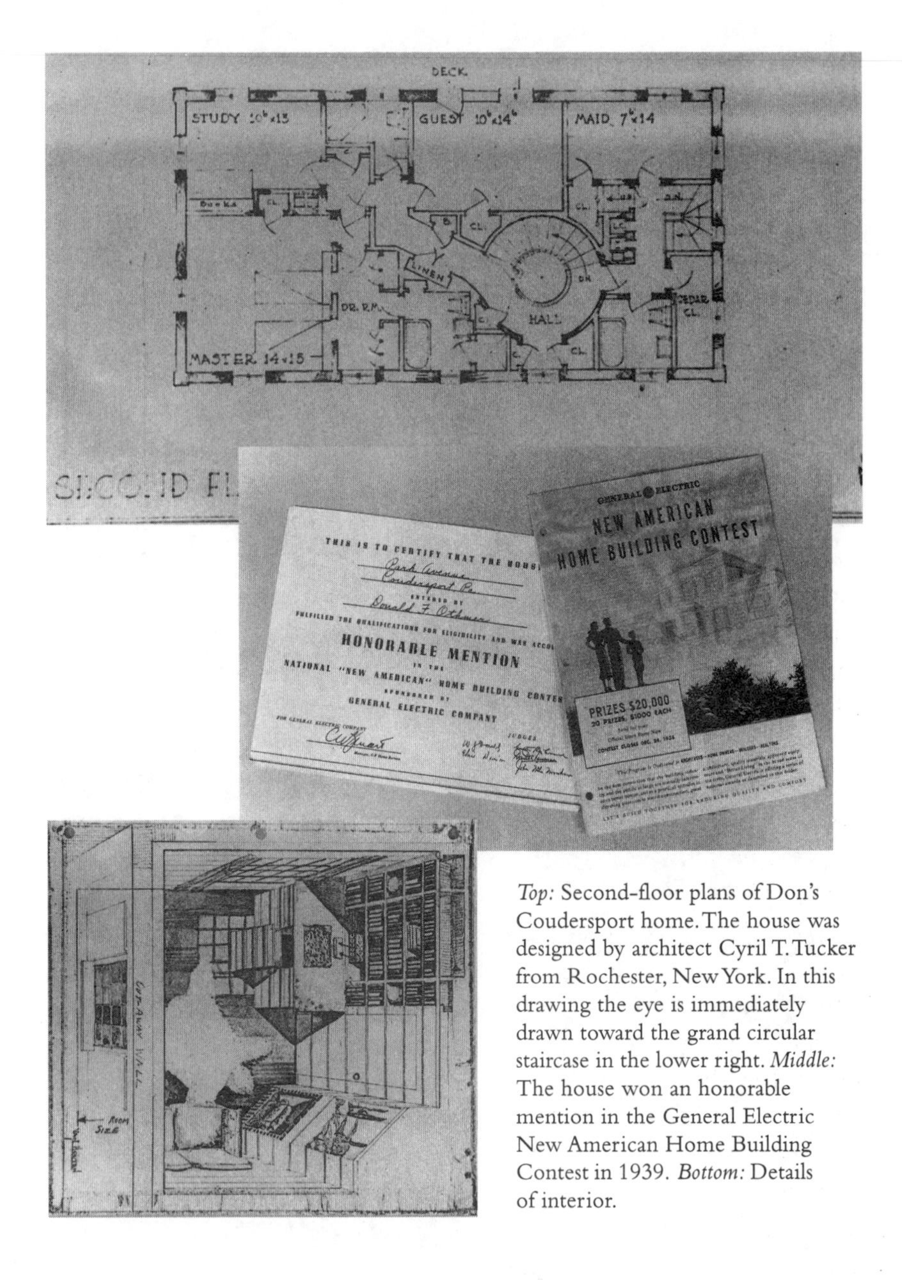

Top: Second-floor plans of Don's Coudersport home. The house was designed by architect Cyril T. Tucker from Rochester, New York. In this drawing the eye is immediately drawn toward the grand circular staircase in the lower right. *Middle:* The house won an honorable mention in the General Electric New American Home Building Contest in 1939. *Bottom:* Details of interior.

holiday cards—the earliest signed by Don and dated 1929. Indeed, Don and Mid's first holiday greeting is found in this informal collection. Held in place with a red ribbon, a transparent piece of paper covers a photo from Don and Mid's wedding with the inscribed greeting: "Our wedding bell rings for *You* for Christmas and for 1951. *Cheers!* Mid and Don Othmer."

Beginning in 1960—and continuing through 1986—Don and Mid

began producing their yearly holiday greetings on 6 × 6-inch ceramic tiles. The tiles, manufactured by the Iwao Jiki Kogyo Company in Japan, were designed by Don and Mid. A holiday message was inscribed on the back, along with the signature salutation "Cheers! Mid and Don Othmer." In virtually every personal and professional file of the archives can be found letters of thanks for these annual greetings. Recipients added to their collections from year to year; some framed them, others mounted them around fireplaces, and still others used them as ashtrays and accent pieces in their homes.

This streak of creative talent, however, was not new to Don; his creativity knew no bounds even in his early years. At age seventeen Don designed a "light fixture for my desk which swings so that it is over my drawing table." His brief journal entries chronicle how he turned the base on a lathe, assembled the parts, and eventually wired it for use. Sixteen years later he would be hard at work on a custom-built home in the Allegheny Mountains near Coudersport, Pennsylvania. The initial plan was for a two-room log cabin. The cabin ultimately evolved into an "eight-room house, three baths, two lavatories, full attic and basement, and as many modern conveniences as could be crammed therein."

The dream house, first sketched out by Don, was designed by architect Cyril T. Tucker in Rochester, New York. Photographs documenting the building of the house and its interior are part of the Othmer archives. Delighted with his magnificent home, Don entered it in the General Electric New American Home Building Contest in 1938. The purpose of the contest was to focus attention on modernized homes that had been built with a strict adherence to "sound specifications." Armed with photographs, blueprints, and an application form that listed all the modern conveniences installed in the home, Don sent in his entry form. The reward for his efforts was an Honorable Mention for his luxurious Coudersport home.

Don was indeed a man of practicality. His personal and professional correspondence was arranged in three major categories—Polytechnic association, geographic location, and individual—which enabled the quick and efficient handling of correspondence. Don's meticulous and practical nature was echoed in the performance of his secretaries, Betty Lou Hornick and Louise Papavero. These women were his professional lifeline when Don was working from his office and most certainly as he traveled abroad. Throughout Don's long career these women typed, phoned, cabled, and greeted countless colleagues and clients for his consulting work. Betty worked with Don throughout the 1940s and into the mid-1960s; Louise took over after Betty retired and worked with Don unceasingly—even providing occasional typing services by mail after she retired to Delray Beach, Florida, in the mid-1970s. "I miss the stimulation of working in an academic atmosphere," she wrote to Don in 1981. "In fact, I miss working!"

Without a secretary Don devised a very practical solution to answering correspondence. He knew, of course, that typing responses could be very time consuming. To save time, Don photocopied the letter he received, wrote his response on the copy, and added a stamped message of explanation:

PLEASE EXCUSE
This informal answer saves delay
necessary for writing a letter.
DONALD F. OTHMER

The capable secretarial assistance he received, enhanced by his own creativity, enabled Don to achieve so much in his seven-decade career at Polytechnic and, simultaneously, as a much-in-demand consultant engineer. "I have been fortunate in . . . always being able to do just what I wanted to do," said Donald Othmer in his oral history. "By and large I have enjoyed all of my life's work." Indeed, it would be hard to believe that without such profound love for his work, Don would have achieved as much as he ultimately did. In 1998 *Chemical & Engineering News* named Don one of the seventy-five most influential contributors to the chemical enterprise, but this was only one of many great honors.

Don received eighteen prestigious awards over the length of his career, among them the Tyler Award and Founders Award from the American Institute of Chemical Engineers (1958 and 1991, respectively), the Murphree Exxon Award from the American Chemical Society and the Perkin Medal from the Society of Chemical Industry (both in 1978), the Chemical Pioneers Award from the American Institute of Chemists (1977), and the Golden Jubilee and Hall of Fame Awards from the Illinois Institute of Technology (1975 and 1981, respectively). Don also received honorary doctorates from the University of Nebraska (1962), Polytechnic University (1977), and New Jersey Institute of Technology (1978). In addition to his many honors, in 1961, Don was named Polytechnic's very first Distinguished Professor.

In letters of recommendation written for his students, Don often wrote that the accomplishments of his students were a source of great pride for him. In an address at a Conversazione of the Center for the History of

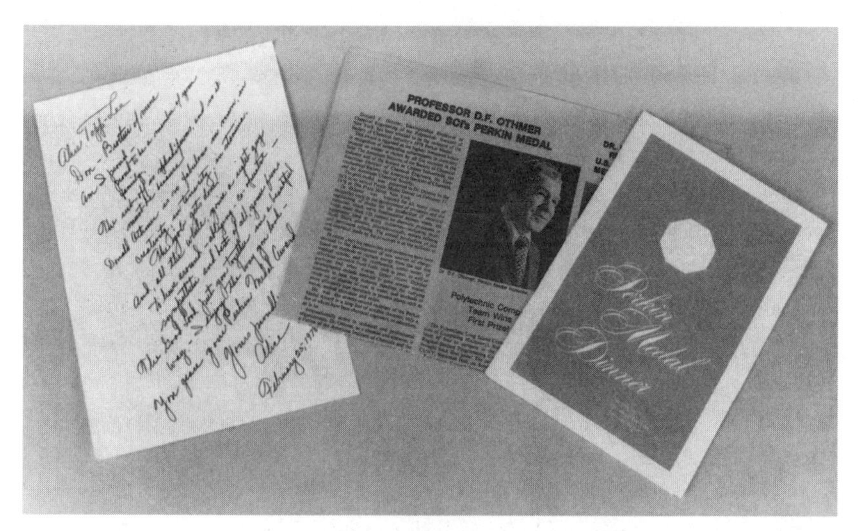

Don was the honored recipient of the Perkin Medal in 1978. The note is from Alice Topp-Lee, Don's sister-in-law. "Don—Brother of mine. Am I proud—proud to be a member of your family. The rub-off is splendiferous, and so it must be because Donald Othmer is so fabulous in vision, in creativity, in tenacity, in stamina. The job gets done! And, all the while, you're a right guy to have around—obliging, cooperative—sympathetic and best of all, you're fun. The Good Lord put you together in a beautiful way—I enjoy the way you look—You grace your Perkin Medal Award. Yours proudly, Alice. February 25, 1978."

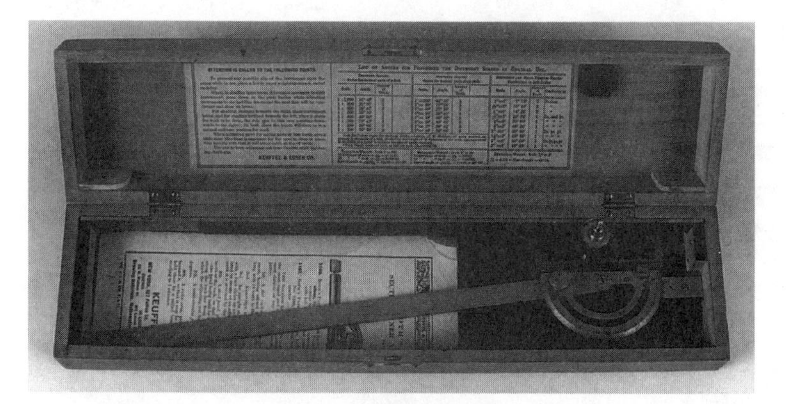

One of the tools of Don's livelihood: Both's Patent Section Liner and Scale Divider, manufactured in German silver by Keuffel & Esser Company, New York. The instrument is in its original wooden case along with the original instructions.

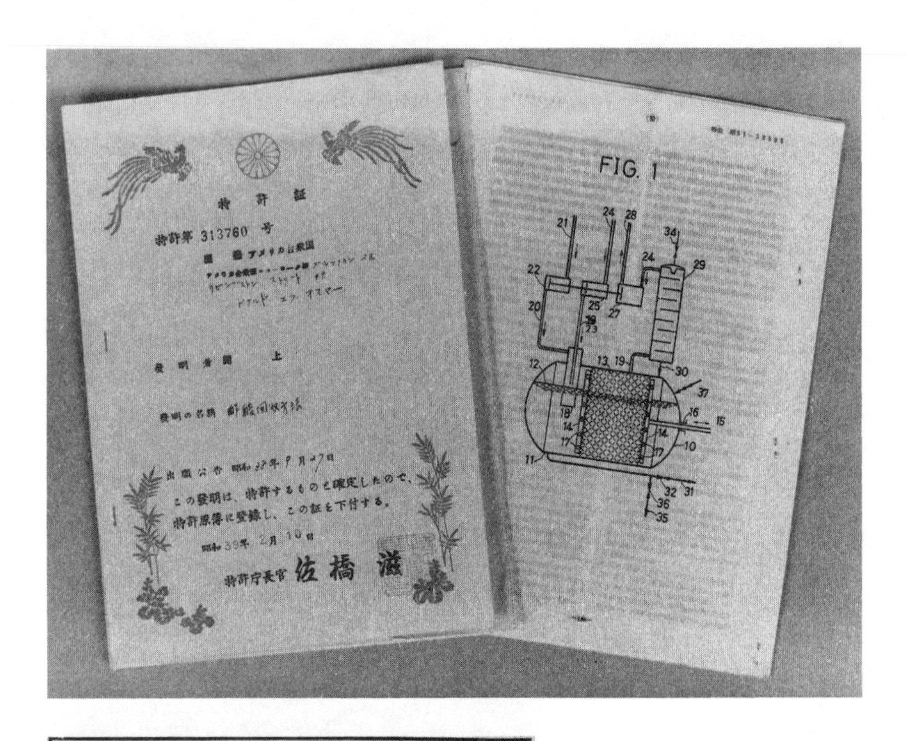

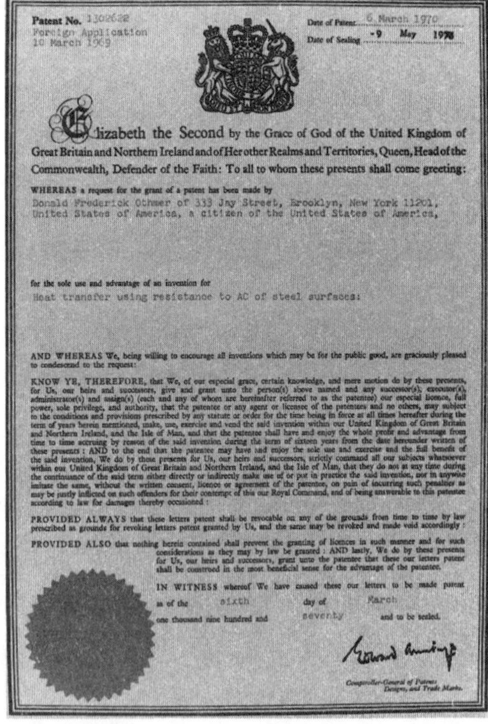

Patent No. 1302622
Foreign Application
10 March 1969

Date of Patent 6 March 1970
Date of Sealing -9 May 197[illegible]

Elizabeth the Second by the Grace of God of the United Kingdom of Great Britain and Northern Ireland and of Her other Realms and Territories, Queen, Head of the Commonwealth, Defender of the Faith: To all to whom these presents shall come greeting:

WHEREAS a request for the grant of a patent has been made by
Donald Frederick Othmer of 333 Jay Street, Brooklyn, New York 11201, United States of America, a citizen of the United States of America,

for the sole use and advantage of an invention for
Heat transfer using resistance to AC of steel surfaces:

AND WHEREAS We, being willing to encourage all inventions which may be for the public good, are graciously pleased to condescend to the request:

KNOW YE, THEREFORE, that We, of our especial grace, certain knowledge, and mere motion do by these presents, for Us, our heirs and successors, give and grant unto the person(s) above named and any successor(s), executor(s), administrator(s) and assign(s) (each and any of whom are hereinafter referred to as the patentee) our especial licence, full power, sole privilege, and authority, that the patentee or any agent or licensee of the patentee and no others, may subject to the conditions and provisions prescribed by any statute or order for the time being in force at all times hereafter during the term of years herein mentioned, make, use, exercise and vend the said invention within our United Kingdom of Great Britain and Northern Ireland, and the Isle of Man, and that the patentee shall have and enjoy the whole profit and advantage from time to time accruing by reason of the said invention during the term of sixteen years from the date hereunder written of these presents: AND to the end that the patentee may have and enjoy the sole use and exercise and the full benefit of the said invention, We do by these presents for Us, our heirs and successors, strictly command all our subjects whatsoever within our United Kingdom of Great Britain and Northern Ireland, and the Isle of Man, that they do not at any time during the continuance of the said term either directly or indirectly make use of or put in practice the said invention, nor in anywise imitate the same, without the written consent, licence or agreement of the patentee, on pain of incurring such penalties as may be justly inflicted on such offenders for their contempt of this our Royal Command, and of being answerable to the patentee according to law for damages thereby occasioned:

PROVIDED ALWAYS that these letters patent shall be revocable on any of the grounds from time to time by law prescribed as grounds for revoking letters patent granted by Us, and the same may be revoked and made void accordingly:

PROVIDED ALSO that nothing herein contained shall prevent the granting of licences in such manner and for such considerations as they may by law be granted: AND lastly, We do by these presents for Us, our heirs and successors, grant unto the patentee that these our letters patent shall be construed in the most beneficial sense for the advantage of the patentee.

IN WITNESS whereof We have caused these our letters to be made patent as of the sixth day of March one thousand nine hundred and seventy and to be sealed.

Comptroller-General of Patents
Designs, and Trade Marks.

Above: Cover page with detail. Japan, Patent 313,760. Process for recovering acetic acid, 10 February 1964. *Left:* Cover page. United Kingdom, Patent 1,302,622. Heat transfer using resistance to AC of steel surfaces, 10 January 1973.

Chemistry in November 1987, Don reminisced about his professorial experience:

> The greatest rewards of my life have come from working with students both full and part time, watching them mature and advising them with their thesis problems—a couple of hundred at least over these years. Thesis problems were always their own selection. Sometimes if their company allowed, as often was the case and the problem was in the company's field, the research was done in the company's laboratories.
>
> What a diversity of research resulted! But this has opened to me the opportunity for insight into so many parts of our industry.
>
> One man, interested in doctoral research in a field of his company's work, had difficulty in getting permission. His boss said, "We can't let Professor Othmer know our development." That was settled easily. I became a consultant with the usual secrecy obligation on this particular problem for $1 per year, during the two or three years involved. But never came the dollar!"

However much Don enjoyed interacting with his students, it is clear that they loved and appreciated him as well. In 1974, for his seventieth birthday, Don's former students organized a celebration to honor their professor and mentor. Reminiscences were collected for "Doc," as they called him, and a color portrait was presented to him (later donated to Polytechnic University). One of his students, Takeshi Utsumi, wrote from Tokyo:

> In various aspects, I owe you personally many, many things. Reminiscing the days of hard study and work at the school, all of my memories are now strangely turned to joyous ones. I appreciate you for your giving me numerous experiences, which I can recall now with my pleasures.

Another student drew a series of eight cartoons for his former professor. One shows a rocket shooting across the void of space, with graphical data connecting it to the earth far below. A voice emits from the rocket: "Say, Houston control, don't you think that Othmer is extrapolating vapor equilibrium data too far?" Another shows Don pushing a wheelbarrow of gold toward a barn with his Othmer still percolating in the background as he says, "My still sure makes good w[h]iskey."

All Don's awards—prestigious, affectionate, and humorous—as well as their documentation, nominations, and programs are part of this extensive collection.

"*You* can have anything in the world you want to," Don wrote in his 1921 journal, "if you are willing to pay the *price*." Don's price was a lifetime of tenacious hard work, but his reward was the joy, fulfillment, and respect

Right: Frontispiece. Vannuccio Biringuccio's posthumously published *Pirotechnia* (Venice, 1550). *Below:* Frontispiece. Hieronymus Brunschwig's *Liber de Arte Distillandi de Compositis* (Strassburg, 1512). Both from the Donald F. and Mildred Topp Othmer Library of Chemical History Rare Book Collection.

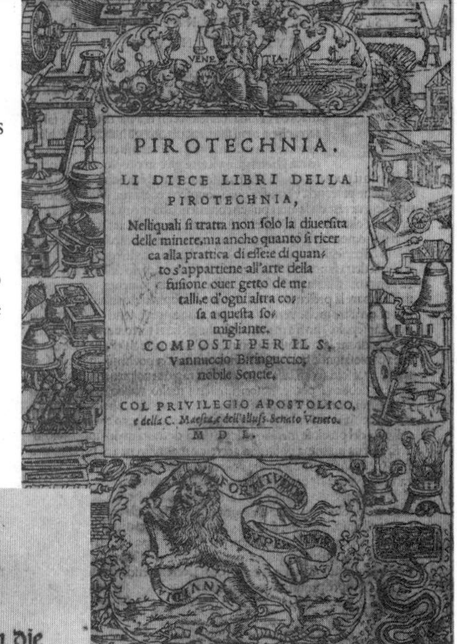

VENE TIA

PIROTECHNIA.

LI DIECE LIBRI DELLA PIROTECHNIA,

Nelliquali si tratta non solo la diuersita delle minere, ma ancho quanto si ricerca alla prattica di esse: e di quanto s'appartiene all'arte della fusione ouer getto de metalli, e d'ogni altra cosa a questa somigliante.

COMPOSTI PER IL S. Vannuccio Biringuccio, nobile Senese.

COL PRIVILEGIO APOSTOLICO, e della C. Maestà, e dell'illuss. Senato Veneto.

M D L.

Conuentus S. Rtbj Salisburg.

Liber de arte Distil landi de Compositis.

Das büch der waren kunst zu distillieren die Composita vñ simplicia/ vnd dz Büch thesaurus pauperū/ Ein schatz d armē genāt Micariū/ die brösamlin gefallen võ dē büchern d Artzny/ vnd durch Experimēt võ mir Jheronimo brūschwick vff geclubt vñ geoffenbart zū trost denē die es begerē.

Das. XXXVII. Capitel.

versüch eß ob wasser dar by sy/ dā so thū dē wein in eī glaß/ setz dā in Balneū marie vnd distillier das fier mal. Vnd wan das geschehen ist/ dan nym ouch diß stück.

Lignum aloes crude
Cardemomi
Cubebe Ana. ʒ. i.
Cinamomi
Nucis muscati
Macis
Zinziber albi
Piperlongi
Grana paradisi

muscati

mit löcher in Balneum marie/ vnnd laß ston. vi. wochen in der stetten hitz das es nit erkalt/ vnnd laß ouch ein stund sieden so würt es goltfar/ das wasser behalt sunder das ist das Aurū potabile. Dar nach nym võ dē erstē Quinta essentia gedistilliert eī maß/ thū darzū yn dise recept wie her nach folgt.

Distillation apparatus (detail from *Liber de Arte Distillandi*, Hieronymus Brunschwig [Strassburg, 1512]).

from a profession very few individuals are fortunate enough to experience. It is a legacy seen not only in his accomplishments but also in the memories of his friends, in the successes of his students, and in the library that bears his name.

For Further Reading

"Contributors to the Chemical Enterprise: C&EN's Top 75." *Chemical & Engineering News 76:2* (1998), pp. 171–185.

Kirk, Raymond E., and Donald F. Othmer, eds. Introduction to *Encyclopedia of Chemical Technology,* vol. 1. New York: Interscience, 1947.

Othmer, Donald F., to General Electric Home Bureau. 21 December 1938. Othmer Archives. Philadelphia: Chemical Heritage Foundation.

Othmer, Donald, and Mildred. Dear Folks Letters. Othmer Archives. Philadelphia: Chemical Heritage Foundation.

Othmer, Donald, interview by James J. Bohning. *Oral History 35,* 1986–1987. Philadelphia: Chemical Heritage Foundation.

Othmer, Donald. A Chemical Engineer Observes Progress. Paper presented at a Conversazione of the Center for the History of Chemistry. 17 November 1987, Philadelphia.

Othmer, Donald. Personal Diaries, 1921–1924. Othmer Archives. Philadelphia: Chemical Heritage Foundation.

Papavero, Louise, to Donald F. Othmer. 17 February 1981. Othmer Archives. Philadelphia: Chemical Heritage Foundation.

Utsumi, Takeshi, to Donald Othmer. 9 May 1974. Othmer Archives. Philadelphia: Chemical Heritage Foundation.

Yeh, Percy P. C., to Donald Othmer. 25 October 1960. Othmer Archives. Philadelphia: Chemical Heritage Foundation.

Section 5

Midon's Institutional Beneficiaries

Brooklyn Botanic Garden

Founded in 1910 on an ash dump, the Brooklyn Botanic Garden is one of the greatest success stories in urban land reclamation. It is an oasis of great tranquility and beauty in the middle of a city of 7.5 million people and is visited by over 600,000 people each year. They come to learn about plants and gardening, attend events, or enjoy a leisurely stroll surrounded by the beauty of nature.

The garden is actually composed of many beautiful gardens. The Japanese Hill-and-Pond Garden, the Cranford Rose Garden, the Fragrance Garden, the Shakespeare Garden, and the Children's Garden are just a few of the most beloved sanctuaries. Each has its own history, unique features, and special attractions, and each is designed to offer a memorable experience that is aesthetic, recreational, and educational.

The garden's mission statement embraces the traditional horticultural, educational, and scientific goals of botanic gardens, but also recognizes its unique focus as an *urban* garden, with special obligations to its immediate constituency from diverse economic and cultural backgrounds. Throughout its eighty-eight-year history, the garden has striven to enhance the quality of daily life in the Brooklyn community through teaching the cultivation and appreciation of plants.

From its inception, the garden's priorities have been to serve not only its urban community but also its worldwide membership, by horticultural example and by popular education, all buttressed by a solid foundation of scientific research. The garden has dedicated itself to the vital task of raising public awareness of threats to the natural world by creating an understanding of the essential role plants play in our lives and the role people must play in protecting and restoring the environment.

The garden's publications reach thousands of avid gardeners around the world. The *21st Century Gardening Series* handbooks explore the frontiers of environmental gardening. The *Plants & Gardens* newsletter is distributed quarterly to the garden's 19,000 members. The *Gardener's Desk Reference*, published in 1999, provides a breadth and quality of information never before available in one authoritative volume.

The garden has many wonderful educational programs for children and adults. Its children's educational programs reach 100,000 children each year. The adult education department offers classes, tours, and workshops from introductory to advanced levels, taught by leading horticulturists and artists. A Certificate of Horticulture program is offered to train people for careers in the "Green Industry."

The garden's community horticulture initiatives promote greening programs, serving an estimated 35,000 Brooklyn residents annually. Over forty groups, including school groups, have implemented long-term community gardens with the assistance of the Brooklyn Botanic Garden, including technical training, plant giveaways, and community horticultural news. On the scientific front horticultural taxonomy is the core of the garden's research activity. The garden has also embarked on a twenty-year urban biodiversity project to inventory the plants of the New York metropolitan area.

The beautiful gardens, as well as the pioneering education programs, have made a lasting impression on a great many people's lives, including the lives of Donald and Mildred Othmer. The Othmers must have developed a deep appreciation of the Brooklyn Botanic Garden and understood its importance to the Brooklyn community.

Mildred Othmer's extraordinarily generous bequest will be used to strengthen the garden's endowment and to develop a master plan for the optimal use of the garden's physical resources, a vital project as the garden approaches its ninetieth year. This stunning bequest will enhance not only the garden but also the Brooklyn community that Mrs. Othmer loved.

The Brooklyn Historical Society

The Brooklyn Historical Society is an extraordinary urban resource center. It was established in 1863 as the Long Island Historical Society, when Brooklyn, an independent city, was the principal economic, social, and political center of Long Island. Renamed in 1985 to better reflect its focus, The Brooklyn Historical Society serves as a museum, library, archive, and educational center dedicated to encouraging the exploration and appreciation of Brooklyn's rich history and heritage. The Brooklyn Historical Society fulfills its mission by collecting, preserving, and making available important materials representative of Brooklyn's diverse peoples and cultures, both past and present.

The Othmer bequest provides the Society with a solid underpinning as it prepares to meet the challenges of the new millennium. The bequest supports the refurbishment and enhancement of the library and collections. The Brooklyn Historical Society's forthcoming renovation of its landmark structure on Pierrepont Street will transform the first-rate Donald F. and Mildred Topp Othmer Library into a state-of-the-art research facility and will establish an exceptional educational resource in the Brooklyn History Discovery Center. The Society will long remember both Othmers with fondness and gratitude.

Chemical Heritage Foundation

The Chemical Heritage Foundation (CHF), established by the American Chemical Society and the American Institute of Chemical Engineers, had its origins in 1982. CHF's mission is to foster the heritage of the chemical sciences and the chemical process industries. Twenty-two professional organizations support CHF in carrying out its mission.

The foundation has three primary goals: strengthening public understanding of the chemical sciences and technologies; increasing the flow of the ablest students into the chemical sciences and chemical process industries; and instilling in chemical scientists and engineers a greater pride in their heritage and their contributions to society. Toward these ends CHF provides a wide range of quality programs—public education, teachers' guides, library and archival activities, publications, public events, and traveling exhibits. CHF also sponsors a visiting scholars program and supports a scholarly community dedicated to producing research on the chemical and molecular sciences, technologies, and industries in the modern age. Scholarships and fellowships encourage research in business history; history of science, technology, and medicine; science and technology studies; museums and public understanding; economic history; and chemical and historical education. Secondary-level teachers, science writers, and doctoral students are also encouraged to take advantage of the scholarships offered at CHF.

CHF fulfills its mission in part through the operations of the Arnold and Mabel Beckman Center for the History of Chemistry and the Donald F. and Mildred Topp Othmer Library of Chemical History. The library was founded in 1988 as a specialized research library to encompass the history of the chemical sciences and the history of the chemical industries. The Othmer Library not only serves as a resource for the foundation's services to the chemical community, but is also used directly by the worldwide chemical community.

The Othmer Library's holdings include some sixty thousand volumes of monographs, reference works, rare books, and multimedia formats in fifteen languages published from 1512 to the present. The collection also includes more than 1,450 journal titles, some of which span almost two hundred years. The Othmer Library houses nearly one thousand historical textbooks as well as a collection of rare and unusual works, including De Fourcy's *Table du produit des affinites chymiques* (1773), which is a precursor of the periodic table; Paracelsus's *Opera omnia* (1658); and Agricola's *De re metallica* (1657). The collection also holds a Hoover translation of *De re metallica* (1912) personally inscribed by both Herbert C. and Lou H. Hoover to Donald Othmer.

The Othmer Library also houses archival, pictorial, and artifact collections. Archives of notable individuals include those of Paul Bartlett, Paul J. Flory, Carl S. "Speed" Marvel, and Donald Othmer. The Pictorial Collection comprises over ten thousand images in all photographic formats, including 35-millimeter slides, negatives, and prints, as well as older formats such as lantern slides. Some two hundred framed pieces of artwork range from oil portraits to watercolors, drawings, pastels, etchings, and lithographs. Archival artifacts include stockings and other products from the early days of nylon, as well as a rapidly growing collection of instruments and apparatus of historical significance.

The Othmer Library also houses an ever-growing collection of oral histories, representing a diverse range of eminent individuals from many fields within the chemical sciences and chemical process industries. Carl Djerassi, Stephanie Kwolek, Donald Othmer, and Linus Pauling are among those with whom oral history interviews have been conducted.

As a commemoration of the life and achievements of Donald Othmer, CHF has established the Othmer Gold Medal. The medal celebrates the richness and enduring strength of our chemical and wider scientific heritage; recognizes the extraordinary achievement that shaped this heritage; and inspires us to draw on this vital resource as we embrace the challenges of tomorrow. It acknowledges multifaceted individuals who, like Donald Othmer, have made enduring contributions to our chemical and scientific heritage through exceptional activity in the areas of innovation, entrepreneurship, research, education, public understanding, legislation, or philanthropy. Ralph Landau, Mary Lowe Good, and P. Roy Vagelos have each received the medal.

For the creation and endowment of the Othmer Library of Chemical History, which embodies their love and support of scientific technological achievement, CHF will forever be grateful to Don and Mid Othmer.

Long Island College Hospital

When first encountered, the word *college* in the name of Long Island College Hospital (LICH) may seem unusual. The explanation is that when LICH was founded in a Brooklyn Heights mansion in 1858, it was a medical school as well as a hospital. In 1860 LICH became the first U.S. medical school to make bedside teaching a standard part of its medical curriculum, establishing an approach that was subsequently adopted throughout the country.

From this auspicious start LICH continued to be a pioneer in medical education and practice well into the twentieth century. "Firsts" achieved by its early faculty include the introduction of the stethoscope (Austin Flint, M.D.); the discovery of the paraurethral glands (Alexander J. C. Skene, M.D.); the first use of skin grafts (Frank H. Hamilton, M.D.); and early use of anesthesia (John C. Dalton, M.D.). The precursor of today's American Society of Anesthesiology was organized at LICH. Hoagland Laboratory, the first hospital-based bacteriology lab in the United States, completed major work on pasteurization of milk, treatment of tuberculosis, and treatment of diphtheria.

In 1930 the Long Island College of Medicine was incorporated as a separate institution to assume the medical education functions of Long Island College Hospital, with LICH as its hospital affiliate. In 1954 the College of Medicine became part of the State University of New York Health Science Center at Brooklyn, commonly known as Downstate Medical Center. Today LICH remains its primary teaching affiliate, offering training programs for resident physicians in more than twenty medical specialties. The LICH School of Nursing, established in 1883, continues as part of the hospital.

LICH currently has 516 beds licensed to provide general medical and surgical care, medical and cardiac intensive care, maternity, nursery, neonatal intensive care, pediatrics, pediatric intensive care, psychiatry, alcohol detoxification, and acute rehabilitation. All tertiary-level services are available on-site, except open-heart surgery and organ transplants. LICH is also nationally recognized for clinical excellence in many areas, including nephrology, urology, and asthma care, and is the home of the Stanley S. Lamm Institute for Child Neurology and Developmental Medicine. LICH's New York Center for Bloodless Medicine and Surgery has become the preeminent program for bloodless care in New York City.

The hospital has 700 physicians and dentists and employs a total staff of approximately 2,800. Its physicians, staff, and administration take pride in combining the teaching and research features of a major medical center with the personal, caring approach of a community-centered hospital.

In May 1998 Long Island College Hospital became the newest member of Continuum Health Partners, Inc., joining the Beth Israel Health Care System and St. Luke's-Roosevelt Hospital Center. With the addition of LICH, Continuum Health Partners offers outstanding primary and specialty care in three large and complementary geographic areas, making the partnership a powerful presence in the New York metropolitan area. This collective use of clinical resources provides LICH patients with state-of-the-art medical services, as well as a seamless referral system that includes a full spectrum of specialty services.

The funds generated from the Othmer endowment will be used throughout Long Island College Hospital. A large portion of the endowment will go toward the capital needs of the institution, including repairs to the physical plant and crucial technological upgrades, such as the computerization of the hospital and the acquisition of new state-of-the-art equipment for the operating rooms. Nursing units and programs will also be upgraded to provide better quality care in a warm and caring environment. Through the benevolence of Donald and Mildred Othmer, LICH will be able to continue its long tradition of excellence in medical training and patient care well into the next century.

Planned Parenthood of New York City

When Margaret Sanger opened the nation's first birth-control clinic in 1915, she sought to provide poor women with hope for the future. Sanger understood that control over fertility meant healthier children born out of choice and nothing less than escape from poverty for many women and their families.

Since 1916 Planned Parenthood of New York City (PPNYC) has worked to fulfill Margaret Sanger's vision through a threefold mission of clinical services, education, and advocacy. Each year PPNYC provides comprehensive and affordable reproductive health care services through nearly fifty thousand visits to family planning centers in the Bronx, Brooklyn, and lower Manhattan. For many lower-income women these services often provide the sole source of compassionate health care and counseling. PPNYC also provides sexuality education and educational training to young people, parents, teachers, and other youth advocates through a variety of programs and groundbreaking, community-based teen pregnancy prevention initiatives. Through the Margaret Sanger Center International, PPNYC helps governmental and nongovernmental organizations develop and enrich sexuality education and family planning programs around the world. In addition, advocacy and public policy work at all levels of government ensures that all women, regardless of age or income, have full access to reproductive health services and counseling.

Throughout PPNYC's history volunteers have played a central role, providing support in nearly every aspect of the agency's operations. Mildred Topp Othmer was one such volunteer, serving as a member of the board of directors and working for many years in the thrift shop. The incredibly generous and thoughtful legacy that she and Donald have left to this agency is deeply appreciated in countless ways but perhaps most especially as a special tribute to PPNYC's volunteers and a spirit of personal commitment and service.

Following Mildred and Donald's wishes, their bequest has been permanently restricted for endowment purposes and will provide PPNYC with a new level of financial stability and security. As a nonprofit agency PPNYC depends on private support, and fluctuations in annual fundraising, uncertain levels of government support, and the needs of a largely uninsured client base severely affect the cash flow situation. While the need for public and private support remains as urgent as ever, the endowment fund created by the Othmers will substantially increase the impact of all contributions made to the agency and enable PPNYC to bring its best ideas to fruition, finding the most creative and effective ways to improve the health, rights, and lives of those who count on it.

As we honor Mildred and Donald's memory, we embrace a future full of challenge and opportunity. On behalf of PPNYC's board of directors, clients, volunteers, and staff, we are profoundly grateful for their legacy of caring and support that will help advance our historic role as a trailblazer in clinical services, research, education, training, and advocacy.

Alexander C. Sanger, President

Plymouth Church of the Pilgrims

From the start Plymouth Church of the Pilgrims was truly a congregational church with a dynamic ministry based on the close collegial cooperation of ordained and lay ministers. The church was founded in 1847 by a group of Congregational laymen and women who then called Henry Ward Beecher to be the church's first minister. Beecher's oratory gave Plymouth Church a national pulpit and ministry.

The issue of slavery brought the church to a national standing as Beecher and his congregation sought to end slavery. Abraham Lincoln was invited to speak on the issue at Plymouth as Beecher and the congregation made Plymouth into a place known as the "Grand Central Depot of the Underground Railroad." The environment was sufficiently hostile that the church and Beecher were picketed by pro-slavery partisans in the 1850s.

The opposition did not hold them back. Members of Plymouth Church sent boxes of Sharp's rifles to the anti-slavery forces in the Kansas-Nebraska border conflict. Because they were sent in boxes labeled "Bibles," those rifles became known as "Beecher's Bibles," and Plymouth Church became known as the "Church of the Holy Rifles."

After Lincoln signed the Emancipation Proclamation in 1863, the national debate changed from whether black people could be enslaved to whether black men should be permitted to vote. Plymouth Church and its minister supported everyone's right to vote—regardless of race or sex.

After the Fifteenth Amendment established that the right to vote would not be abridged because of race, the question became whether women would be permitted to vote. In the pre–Civil War era, Northern churches had opposed slavery on biblical grounds, citing such passages as Paul's letter to the Galatians as support: "There is neither Jew nor Greek, slave nor free, male nor female, for you are all one in Christ Jesus" (Galatians 3:28). Obviously, this passage could be used to support the right of women to have equality. Plymouth Church and Beecher then came to support women's suffrage.

The underlying conviction that propelled Plymouth Church of the Pilgrims into national issues was a deep and energizing faith in Jesus Christ. The church's first purpose was evangelism, the bringing of people, especially the unchurched of their day, to become followers of Jesus Christ. It is the same today as it was then. From May 1997 to May 1998 the congregation of Plymouth Church of the Pilgrims in Brooklyn, New York, celebrated the sesquicentennial of the founding of Plymouth Church. During this time the members have reaffirmed the purpose of the church: To bring and bind together followers of Jesus Christ for worship, discipleship, fellowship, and inreach and outreach ministry.

Mildred and Donald Othmer belonged to Plymouth Church for many years, subscribing to its faith and history, contributing to its present years through their participation in the corporate life of the church as members. Plymouth's deep grounding in the Christian faith, its Christ-centeredness, resonated with the strong faith and the values of the Othmers.

The Othmer bequest will be used to further the church's purpose by reaching out into the greater community, demonstrating and offering God's love through Jesus Christ.

Polytechnic University

Polytechnic University, founded in 1854 in Brooklyn, is the nation's second oldest private engineering and science school. Known for many years as "Brooklyn Poly," the university is the metropolitan area's preeminent resource in science and technology.

Polytechnic, with campuses at MetroTech Center in downtown Brooklyn, in Westchester, and on Long Island, has long been famous for its commitment to academic excellence. The school is especially noted for its electrical engineering, microwave engineering, and polymer chemistry departments. Scientists at Polytechnic, for example, helped develop radar during World War II. Herman Mark, considered the "father of polymer chemistry," established the Polymer Research Institute at the university in 1942.

Celebrated alumni include Robert G. Brown, who designed France's first central telephone system; Henry C. Goldmark, who designed the Panama Canal lock system; Leopold H. Just, who helped design virtually every major bridge in New York City; Martin L. Perl, who won the 1995 Nobel Prize in physics; and Gertrude Elion, who won the 1988 Nobel Prize in physiology or medicine. Internationally famous faculty include Ernst Weber, the microwave engineering pioneer; Donald F. Othmer, co-editor of the *Kirk-Othmer Encyclopedia of Chemical Technology*; Rudolph Marcus, who won the 1992 Nobel Prize in chemistry; and, most recently, David and Gregory Chudnovsky, Russian-born mathematicians and brothers who have been honored with nearly every major award in their field.

Today Polytechnic is on the cutting edge of education in telecommunications, information science, and technology management. In1997, the university unveiled the Institute for Mathematics and Advanced Supercomputing under the direction of the Chudnovsky brothers, who are applying state-of-the-art mathematics to innovations across several scientific disciplines.

The university's graduate division also has a strong international reputation. *U.S. News & World Report,* in 1998, ranked Polytechnic's doctoral program in polymers among the top ten in the nation—ahead of Stanford University, Carnegie-Mellon University, and the Massachusetts Institute of Technology. The electrical engineering doctoral program has been rated in the nation's top ten by the American Society of Engineering Education, based on a study by the Conference Board of Associated Research

Councils. Polytechnic is among the nation's most successful institutions in producing science and engineering graduates who go on to earn Ph.D. degrees.

In 1998 Polytechnic was the beneficiary of an extraordinary $175 million bequest from Dr. Othmer and his wife, Mildred Topp Othmer. Dr. Othmer taught chemical engineering at Polytechnic for nearly sixty years. Their gift enables the university to transform itself into a leading science and technological institution. Polytechnic will use the bequest to establish Othmer-endowed scholarships and an Othmer Endowment Fund. The university will also name a building after the Othmers and create an Othmer Institute for Interdisciplinary Studies, crossing disciplinary barriers to embrace changes in education and research, and an Othmer School for the Chemical Industry to nurture ties to the chemical industry and serve as the preferred place from which to hire chemical engineering graduates.

Appendices

REPUBLIK ÖSTERREICH

PATENTURKUNDE

GEMÄSS DEM PATENTGESETZ IST
FÜR DIE IN DER ANGEFÜGTEN PATENTSCHRIFT
BESCHRIEBENE ERFINDUNG
EIN PATENT UNTER DER
NR. 332950
ERTEILT WORDEN.

DEN 25. OKTOBER 1976

ÖSTERREICHISCHES PATENTAMT
PATENTREGISTER

DIE JA SGEBÜHREN
WERDE LJÄHRLICH FÄLLIG AM 15. FEBRUAR

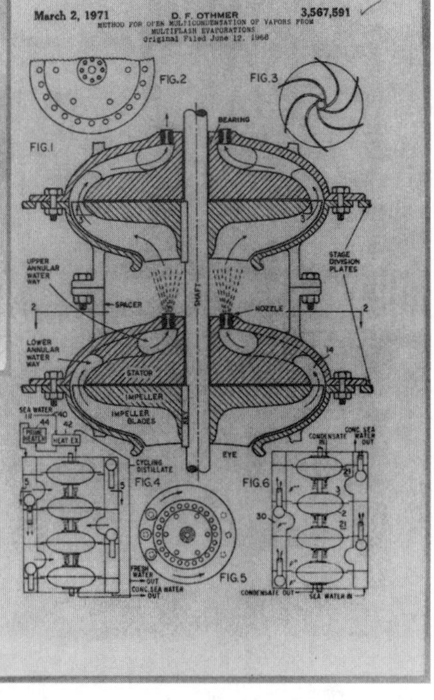

Above: Cover page. Austrian Republic Patent 332,950. System um erwärmen von in einem transportrorh strömenden öl mit einem heizrohr. 9 March 1970. *Right:* Detail. United States Patent 3,616,653. Refrigeration in cycles of freezing and melting. 2 November 1971.

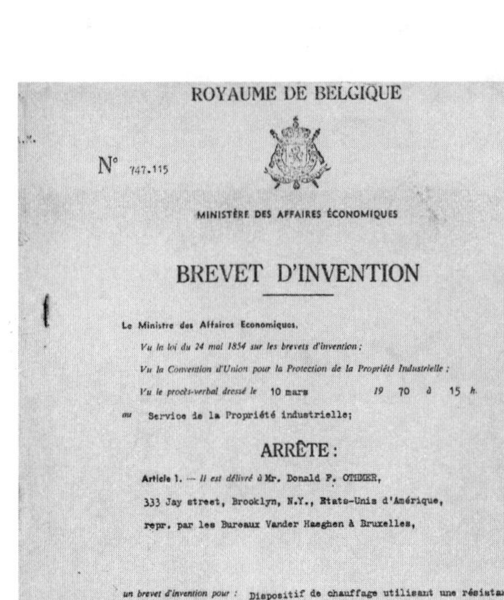

ROYAUME DE BELGIQUE

N° 747.115

MINISTÈRE DES AFFAIRES ÉCONOMIQUES

BREVET D'INVENTION

Le Ministre des Affaires Economiques,

Vu la loi du 24 mai 1854 sur les brevets d'invention ;

Vu la Convention d'Union pour la Protection de la Propriété Industrielle ;

Vu le procès-verbal dressé le 10 mars *19* 70 *à* 15 *h.*

au Service de la Propriété industrielle;

ARRÊTE :

Article 1. — *Il est délivré à* Mr. Donald F. OTHMER,

333 Jay street, Brooklyn, N.Y., Etats-Unis d'Amérique,

repr. par les Bureaux Vander Haeghen à Bruxelles,

un brevet d'invention pour : Dispositif de chauffage utilisant une résistance électrique d'un circuit à courant alternatif,

qu'il déclare avoir fait l'objet d'une demande de brevet déposée aux Etats-Unis d'Amérique le 10 mars 1969 sous le n° 805.718.

Article 2. — *Ce brevet lui est délivré sans examen préalable, à ses risques et périls, sans garantie soit de la réalité, de la nouveauté ou du mérite de l'invention, soit de l'exactitude de la description, et sans préjudice du dr*

Au présent arrêté demeurera joint un des doubles de (mémoire descriptif et éventuellement dessins) signés par de sa demande de brevet.

Bruxelles, le

Pour expédition certifiée conforme :

E. HERMANS,
Conseiller-Chef de Service

Eckenberg
86.117

Cover page. Belgium Patent 747,115. Électrique d'un circuit à courant alternatif. 10 March 1970.

N.

MINISTERO DELL'INDUSTRIA E DEL COMMERCIO

UFFICIO CENTRALE DEI BREVETTI PER INVENZIONI, MODELLI E MARCHI

BREVETTO PER INVENZIONE INDUSTRIALE

Cover page. Italy Patent 735,090. Procedimento per la produzione di acqua pura da acqua di matre e da altre soluzioni, per vaporizzazione istantanea e condensazione. 16 January 1963.

Section 6

Donald Othmer's Patents, Publications, and Awards

Patents

All of the following patents bear the name Donald F. Othmer exclusively, unless otherwise noted.

Number	Title (and Coauthor)	Date
United States		
1,804,745	Dehydrating Aqueous Acetic Acid (with H.T. Clark)	12 May 1931
1,826,302	Formic Acid (with H.T. Clark)	6 Oct. 1931
1,861,841	Dehydrating Aqueous Acetic Acid (with H.T. Clark)	7 June 1932
1,866,719	Removing Tarry Substances from Pyroligneous Liquids	12 July 1932
1,879,445	Crystallizing Salts such as Silver Nitrate from Solutions	27 Sept. 1933
1,884,164	Composition for Removing Coatings such as Varnish, Paint or Lacquer	25 Oct. 1932
1,897 816	Dehydrating Acetic Acid	14 Feb. 1933
1,908,508	Dehydrating Aqueous Acetic Acid with Trichloroethylene (with H. T. Clark)	9 May 1933
1,917,391	Dehydrating Aqueous Acetic Acid	11 July 1933
1,930,146	Formic Acid	10 Oct. 1933
1,939,222	Recovering Ketones such as Acetone from Gaseous Mixtures	12 Dec. 1933
2,000,606	Countercurrent Liquid-Extraction Apparatus Suitable for Extracting Acetic Acid with Ethyl Acetate	7 May 1935
2,028,800	Acetic Acid Concentration	28 Jan. 1936

Number	Title (and Coauthor)	Date
United States (continued)		
2,028,801	Process for Concentrating Acetic Acid	28 Jan. 1936
2,050,234	Concentrating Lower Aliphatic Acids such as Acetic Acid	4 Aug. 1936
2,050,235	Process for Concentrating Acetic Acid	4 Aug. 1936
2,076,184	Concentrating Lower Aliphatic Acids from Their Aqueous Solutions	6 April 1937
2,157,143	Drying Lower Aliphatic Acids such as Acetic Acid	9 May 1939
2,170,834	Dehydrating Acetic, Propionic, Butyric and Other Lower Aliphatic Acids	29 Aug. 1939
2,184,563	Concentrating Acetic Acid	26 Dec. 1939
2,186,617	Separating Acetic Acid from Tarry Components of Pyroligneous Liquid	9 Jan. 1940
2,204,616	Process for Dehydration of Acetic Acid and Other Lower Fatty Acids	18 June 1940
2,220,777	Valveless Chemical Heater Suitable for Various Uses	5 Nov. 1941
2,227,979	Dehydrated Acetic Acid from Pyroligneous Liquors	7 Jan. 1941
2,269,163	Dehydrating Aliphatic Acids such as Acetic Acid	6 Jan. 1942
2,275,802	Treating Azeotropic Condensate (with Robert E. White)	10 March 1942
2,275,862	Extraction of Lower Aliphatic Acids from Aqueous Solutions	10 March 1942
2,278,090	Thiodiglycol (with D. Q. Kern)	31 March 1942
2,290,483	Separation and Recovery of Acetic Acid, etc., as in Treating Mixtures Obtained from Pyroligneous Liquor	21 July 1943
2,395,010	Concentration of Dilute Solutions of Acetic Acid and Their Lower Fatty Acids	19 Feb. 1946
2,406,648	Water Soluble Alkyl Lactates (with S. M. Weisberg, E. G. Stimpson, and J. Greenspan)	27 Aug. 1946
2,482,879	Protein Dispersions and Their Use in Printing Inks (with Alfred F. Schmutzler)	27 Sept. 1949
2,486,974	Recovery of Stearic and Abietic Acids from Hydrogenated Tall Oil (with G. Papps)	1 Nov. 1949
2,525,834	Process of Toner Manufacture (with Alfred F. Schmutzler)	17 Oct. 1950
2,525,835	Process for the Preparation of Coated Pigment Particles (with Alfred F. Schmutzler)	17 Oct. 1950
2,537,101	Process for Manufacture of Wallboard from Lignocellulosic Material (with W. T. Smith)	9 Jan. 1951

Number	Title (and Coauthor)	Date
United States (continued)		
2,552,597	Process of Making a Molded Article from Lignocellulose (with W. R. Smith and E. R. Mellinger)	15 May 1951
2,606,123	Printing Ink (with A.F. Schmutzler)	5 Aug. 1952
2,631,145	Separation of Essential Oils into Component Fractions (with M. B. Jacobs and N. Wishnefsky)	10 March 1953
2,669,522	Molded Products of Lignocellulose and Lignin (with W. R. Smith)	16 Feb. 1954
2,687,556	Wallboard of Compressed Lignocellulosic Material and Chemically Combined Soluble Silicates (with L. G. Ricciardi and W. R. Smith)	31 Aug. 1954
2,692,206	Water-Resistant Molded Cellulose (with E. R. Mellinger and W. R. Smith)	19 Oct. 1954
2,835,572	Method of Making Porous Metal Molds (with Hrant Ishenjian and Leopold Hase)	20 May 1958
2,859,154	Concentration of Acetic Acid and Other Lower Fatty Acids	4 Nov. 1958
2,867,655	Process for Recovering Volatile Aliphatic Acids	6 Jan. 1959
2,878,283	Process for Recovering Acetic Acid	17 May 1959
3,023,254	Dehydrocyclization Process (with Denis K. Huang)	27 Feb. 1962
3,024,170	Process of Azeotropic Distillation of Formic Acid from Acetic Acid (with J. J. Conti)	6 March 1962
3,087,404	Photographic Method (with Leo Hase)	30 April 1963
3,137,995	Ablation Resistant Reaction Propulsion Nozzle (with Walter Brenner)	23 June 1964
3,151,043	Vapor-Liquid Contacting and Mass Transfer (with Robert D. Beattie)	29 Sept. 1964
3,250,081	Process for Freezing Water from Solutions to Make Fresh Water or Concentrated Solutions	10 May 1966
3,288,686	Method for Multi-Flash Evaporation to Obtain Fresh Water from Aqueous Solution	29 Nov. 1966
3,306,346	Method for Cooling Volatile Liquid	28 Feb. 1967
3,325,308	Process for the Refining of Sugar with Two or More Solvents	13 June 1967
3,329,583	Method for Producing Pure Water from Sea Water and Other Solutions by Flash Vaporization and Condensation	4 July 1967
3,377,814	Method for Producing Fresh Water from Slurry of Ice in an Aqueous Liquid	16 April 1968

Number	Title (and Coauthor)	Date
United States (continued)		
3,408,294	Method for Removing Scale-Forming Constituents from Sea Water and Other Solutions Which Form Scale	29 Oct. 1968
3,446,712	Method for Producing Pure Water and Other Solutions by Flash Vaporization and Condensation	27 May 1969
3,567,591	Method for Open Multicondensation of Vapors from Multi-Flash Evaporations	2 March 1971
3,583,895	Evaporation using Vapor-Reheat and Multi-effects	8 June 1971
3,616,653	Refrigeration in Cycles of Freezing and Melting	2 Nov. 1971
3,617,699	A System for Electrically Heating a Fluid Being Transported in a Pipe	2 Nov. 1971
3,692,634	Method for Producing Pure Water from Sea Water and Other Solutions by Flash Vaporization and Condensation	19 Sept. 1972
3,772,187	Sewage Treatment Process	13 Nov. 1973
3,777,117	Electric Heat Generating System	4 Dec. 1973
3,788,476	Sewage Treatment System	29 Jan. 1974
3,793,003	Method for Producing Aluminum Metal Directly from Ore	19 Feb. 1974
3,853,541	Method for Producing Aluminum Metal Directly from Ore	10 Dec. 1974
3,856,508	Method for Producing Aluminum Chloride, Aluminum Metal, and Iron Directly from Ores	24 Dec. 1974
3,859,077	Manufacture of Titanium Chloride Synthetic Rutile and Metallic Iron from Titaniferous Materials Containing Iron	7 Jan. 1975
3,861,904	Method for Producing Aluminum Metal Directly from Ore	21 Jan. 1975
3,928,145	Process for Producing Power, Fresh Water, and Food from the Sea and Sun	23 Dec. 1975
3,938,988	Method for Producing Aluminum Metal from Its Salts	17 Feb. 1976
3,974,398	Wire and Steel Tube as AC Cable	19 Aug. 1976
3,975,617	Pipe Heating by AC in Steel	17 Aug. 1976
3,977,866	Method for Producing Titanium	31 Aug. 1976
3,989,510	Manufacture of Titanium Chloride and Metallic Iron from Titaniferous Materials Containing Iron Oxides	2 Nov. 1976
4,017,421	Wet Combustion Process	12 April 1977
4,116,712	Solvent Refining of Sugar	26 Sept. 1978

Number	Title (and Coauthor)	Date
United States (continued)		
4,151,075	Separation of Components of a Fluid Mixture	24 April 1979
4,251,227	Method for Producing SNG or Syngas from Wet Solid Wastes and Low Grade Fuels	17 Feb. 1981
4,329,156	Desulfurization of Coal	11 May 1982
4,405,343	Methanol Dehydration	20 Sept. 1983
Argentina		
80,002	Procedimento para la Manufacture de Tablero Dura de Pared, Se Serrin y Otros Desperdicios Lignocelulosicos (with L. G. Ricciardi and W. R. White)	15 May 1951
82,969	Procedimento para la Manufacture de Tablero Dura de Pared, Se Serrin y Otros Desperdicios Celulosicos (with L. G. Ricciardi and W. R. White)	16 Jan. 1952
87,666	Procedimento para la Fabricacion de Tableros de Pared y Productos Similars (with L. G. Ricciardi and W. R. White)	19 Dec. 1952
196,067	Una Disposicion Generadora de Calor	30 Nov. 1973
Australia		
149,559	Process for Manufacture of Wallboard and Allied Products (with L. G. Ricciardi and W. R. Smith)	23 April 1953
417,286	Method for Producing Pure Water from Sea Water and Other Solutions by Flash Vaporization and Condensation	11 Feb. 1972
420,462	Method for Producing Pure Water from Sea Water and Other Solutions by Flash Vaporization and Condensation	10 June 1971
420,569	Method of Flash Evaporation of an Original Aqueous Solution to Produce Fresh Water and Having a Scale Removing Step	10 June 1971
Austria		
306,878	System zum elektrischen Erwärmen eines Mediums mit einem Heizrohr	25 April 1973
309,613	System zum Erwärmen eines in einem Transportrohr Transportierten Mediums	27 Aug. 1973
332,950	System zum Erwärmen von in einem Transportrohr Stromenden ol mit einem Heizrohr	25 Oct. 1976

Number	Title (and Coauthor)	Date
Belgium		
747,115	Dispositif de Chauffage Utilisant une Resistance Electrique d'un Circuit à Courant Alternatif	10 March 1970
Canada		
528,355	Dry Molding Process (with W. R. Smith)	31 July 1956
573,073	Method of Making Hard Board (with L. G. Ricciardi)	31 March 1959
952,169	Skin Heating of Steel by A.C.	30 July 1974
955,635	System for Electrically Heating a Fluid Being Transported in a Pipe	1 Oct. 1974
1,035,833	Steel Tube-Wire A.C. Cable	1 Aug. 1978
1,087,129	Wet Combustion	7 Oct. 1980
Egypt		
5,192	Desalination . . .	9 June 1965
5,170	Separation by freezing . . .	21 Dec. 1973
Finland		
24,658	Improvements in Process for Manufacture of Hard Wallboard from Sawdust and Other Cellulosic Waste (with W. R. Smith)	10 Oct. 1950
France		
1,384,939	Desalination . . .	30 Nov. 1964
1,410,993	Desalination . . .	9 Aug. 1965
2,037,832	Dispositif de Transmission de Chaleur Utilisant la Resistance Electrique, en Courant Alternatif, de Surfaces en Acier	21 Dec. 1970
Germany		
1,209,549	Stoffaustauschkolonne (with Robert D. Beattie)	13 Jan. 1961
2,246,652	Verfahren zur Behandlung Wabriger Schmutz-flussigkeiten	28 March 1974
2,636,701	Rohrheizung durch Wechselstrom in Stahl	16 Feb. 1978
Great Britain		
318,633	Formic Acid Concentration (with H. T. Clark)	8 Sept. 1928
354,553	Dehydrating Ethyl Alcohol	29 Aug. 1929
623,991	Dehydration of Acetic Acid	26 May 1949
967,192	Improvements in Vapour-Liquid Contacting and Mass Transfer	29 Aug. 1964

Number	Title (and Coauthor)	Date
Great Britain (continued)		
1,134,130	Method for Multi-Flash Evaporation to Obtain Fresh Water from Aqueous Solution	16 Nov. 1966
1,302,622	Heat Transfer Using Resistance to AC of Steel Surfaces	10 Jan. 1973
Greece		
28,090	Methodos kai Michaniki Diataxis Pros Pliri Psixin Kristallon Dialitikou Mesou ex Enos Dialimatos	11 May 1965
26,705	Methodos Pros Paragogin Kathari Idatos ek Thalassiou Tioutou kai Allon Dialimaton di Akariaias Exatmiseos kai Simiknoseos	25 Sept. 1964
Italy		
735,090	Procedimento per la Produzione di Acqua Pura da Acqua di Mare e Altre Soluzioni per Vaporizzazione Istantanea e Condensazione	15 Dec. 1966
742,592	Procedimento per Raffreddare un Fluido Volatile con Fasi di Evaporazione e Scambio Termico con Altro Fluido in Controcorrente	1967
747,379	Procedimento di Evaporazione Istantanea Multipla per Ottenere Acqua Pura da una Soluzione Acquosa	1 Feb. 1967
898,246	Trasmissione di Calore Mediante Uti Lizzazione della Resistenza Alla Corrente Alternata di Superfici di Acciaio	1 Dec. 1971
Japan		
197,872	Process for Dehydration of Acetic Acid and Other Lower Fatty Acids	28 Jan. 1953
300,726	Process for Recovering Volatile Aliphatic Acid	14 May 1962
418,074	Vapor-Liquid Contacting and Mass Transfer (with R. F. Beattie)	23 Jan. 1964
313,760	Process for Recovering Acetic Acid (with R. F. Beattie)	10 Feb. 1964
315,583	Process for Concentrating Acetic Acid and Other Lower Fatty Acids Using Entrainers for the Acids	21 Sept. 1964
422,726	Dehydrocyclization Process (with D. K. Huang)	8 April 1964

Number	Title (and Coauthor)	Date
Japan (continued)		
821,535	Sewage Treatment Plant Design	16 Dec. 1975
867,539	Sewage Treatment Process	27 Nov. 1976
871,676	Method for Removing Scale-Forming Constituents from Sea Water and Other Solutions Which Form Scale	13 Sept. 1976
Mexico		
48,547	Process for Manufacture of Wallboard and Allied Products (with L. F. Ricciardi and W. R. Smith)	28 May 1951
Portugal		
26,636	Process para o Fabrico de Tabuas Rijas para Revestimento Partindo de Serradura e Outros Desperdicios Celulosicos (with W. R. Smith and E. R. Mellinger)	25 May 1953
South Africa		
7,685	Process for Manufacture of Wallboard and Allied Products (with L. G. Ricciardi and W. R. Smith)	24 Oct. 1949
Spain		
183,790	Procedimiento para la Manufactura de Tablero Duro de Pared, de Serrin y Otros Desperdicios Lignocelulosicos (with Warren R. Smith)	21 May 1948
183,817	Procedimiento para la Manufactura de Tablero Duro de Pared, de Serrin y Otros Desperdicios Celulosicos (with Warren R. Smith)	24 May 1948
183,835	Procedimiento para la Manufactura de Tableros Duros de Pared Partiendode Serrin y Otros Desperdicios Cellulosicos (with W. R. Smith and E. R. Mellinger)	26 May 1948
184,037	Procedimiento para la Manufactura de Tableros de Pared y Productos Similares (with L. G. Ricciardi and W. R. Smith)	10 June 1948
295,108	Desalination . . .	1965
295,788	Desalination . . .	1965

Number	Title (and Coauthor)	Date
Sweden		
170,269	Satt Att Framstalla Harda Och Tata Foremal Genom Pressning Av Lignocellulosa-haltiga Material (with L. G. Ricciardi)	19 Nov. 1959
Switzerland		
587,438	Anordnung zum Erwarmen eines in einem Transportrohr Stromenden Mediums	3 Oct. 1970
542,399	Einrichtung an einem Transportrohr zum Erwarmen eines in diesem Transportierten Mediums	15 Nov. 1973
542,565	Elektrische heizvorrichtung zum Erwarmen eines Mediums mit einem Heizrohr	15 Nov. 1973

Publications

1925

W. L. Badger and Donald F. Othmer. Studies in Evaporator Design, VII—Optimum Cycle for Liquids Which Form Scale. Preprint, AIChE, 23–26 June 1925. *Transactions of the AIChE,* vol. 16, pt. II, 159 (1924).

1928

Donald F. Othmer. Composition of Vapors from Boiling Binary Solutions. *Industrial and Engineering Chemistry, 20,* 743.

Donald F. Othmer and H. B. Coats. Measurement of Surface Temperature. *Industrial and Engineering Chemistry, 20,* 124.

1929

Donald F. Othmer. An All-Glass Evaporator. *Industrial and Engineering Chemistry, 21,* 876.

———. The Condensation of Steam. *Industrial and Engineering Chemistry, 21,* 576.

———. Corrosion Testing Apparatus. *Industrial and Engineering Chemistry, 1,* 209 (Analytical Ed.).

———. Heat Transfer from Steam to Metal. *Engineering (London), 77,* 745.

———. A Heavy-Duty Thermostat. *Industrial and Engineering Chemistry, 1,* 97 (Analytical Ed.).

———. Large-Capacity Laboratory Condensers. *Industrial and Engineering Chemistry, 1,* 153 (Analytical Ed.).

———. Manometer for Determination of Gases in Vapors. *Industrial and Engineering Chemistry, 1,* 46 (Analytical Ed.).

1930

Donald F. Othmer. Accurately Adjustable Gravity Separator. *Chemical and Metallurgical Engineering, 37,* 380.

———. Large Glass Distillation Equipment. *Industrial and Engineering Chemistry, 22,* 322.

———. Trends in Heat Transfer. *Industrial and Engineering Chemistry, 22,* 988.

———. Trends in Heat Transfer. *The Nebraska Blue Print, 29,* 211.

1931

Donald F. Othmer. Glass Temperature and Float Regulators. *Industrial and Engineering Chemistry, 3,* 139 (Analytical Ed.).

———. Sight Feed for Centralizing Control of Distillation Equipment. *Chemical and Metallurgical Engineering, 38,* 415.

1932

Donald F. Othmer. Composition of Vapors from Boiling Binary Solutions. *Industrial and Engineering Chemistry, 4,* 232 (Analytical Ed.).

———. Guarding against Corrosion in Acetic Acid Equipment. *Chemical and Metallurgical Engineering, 39,* 136.

———. Problems in Piping Chemicals Typified by Acetic Acid. *Heating, Piping and Air Conditioning*, *4*, 346.

1933

Donald F. Othmer. Acetic Acid Dehydration. Preprint, AIChE, 12–14 Dec. 1933. *Transactions of the AIChE*, *30*, 299 (1933–34).
———. Dehydrating Aqueous Solutions of Acetic Acid. *Chemical and Metallurgical Engineering*, *40*, 631.
———. Research Yielding Important Advances in Distillation Practice. *Chemical and Metallurgical Engineering*, *40*, 254.

1934

Donald F. Othmer. Acetic Anhydride. *Chemical and Metallurgical Engineering*, *41*, 514.
———. Chemical Engineering as a Career—Design of Equipment. *Poly Men*, *10*, 7.
———. Dehydrating Acetic Acid by Extraction. *Chemical and Metallurgical Engineering*, *41*, 81.
———. Glass for Pilot Plant Construction. *Chemical and Metallurgical Engineering*, *41*, 547.
———. How to Prepare for Chemical Equipment Design. *Chemical and Metallurgical Engineering*, *41*, 187.

1935

Donald F. Othmer. Acetic Acid and a Profit from Wood Distillation. *Chemical and Metallurgical Engineering*, *42*, 356.
———. New Equipment in the Chemical Engineering Laboratory. *Poly Men*, *11*.
———. Science and the Social Sciences. *Poly Men*, *11*, 3.
———. Separation of Water from Acetic Acid by Azeotropic Distillation. *Industrial and Engineering Chemistry*, *27*, 250.

1936

Donald F. Othmer. Vapor Re-Use Process—Separation of Mixtures of Volatile Liquids. *Industrial and Engineering Chemistry*, *28*, 1435.

1940

Paul F. Bruins, Donald F. Othmer, K. A. James, and Martin Berman. Friction of Fluids in Solder-Type Fittings. *Transactions of the AIChE*, *36*, 721.
Donald F. Othmer. Acetic Acid from Wood Distillation. *Chemical and Metallurgical Engineering*, *47*, 349.
———. Correlating Vapor Pressure and Latent Heat Data—A New Plot. *Industrial and Engineering Chemistry*, *32*, 841.
———. New Psychrometric Chart. *Chemical and Metallurgical Engineering*, *47*, 296.
———. Nomographic Plotting Proves Convenient for Vapor Pressures. *Chemical and Metallurgical Engineering*, *47*, 631.
———. Simple Plot for Partial Pressure and Vapor Composition Data for Aqueous Ammonia. *Chemical and Metallurgical Engineering*, *47*, 551.

Donald F. Othmer and J. J. Jacobs, Jr. Anhydrous Sodium Hydroxide—Production by Partial Pressure Evaporation. *Industrial and Engineering Chemistry, 32,* 154.

Donald F. Othmer and Donald Q. Kern. Thiodiglycol—Unit Process and Operations Involved in Its Synthesis from Ethylene Oxide and Hydrogen Sulfide. *Industrial and Engineering Chemistry, 32,* 160.

Donald F. Othmer and T. O. Wentworth. Absolute Alchohol—An Economical Method for its Manufacture. *Industrial and Engineering Chemistry, 32,* 1588.

T. O. Wentworth and Donald F. Othmer. Absolute Alchohol, An Economical Method for its Manufacture. *Transactions of the AIChE, 36,* 785.

1941

Donald F. Othmer. Azeotropic Distillation for Dehydrating Acetic Acid. *Chemical and Metallurgical Engineering, 48,* 91.

———. Cheaper Absolute Alcohol—Unique Method Based on Use of Ethyl Ether as the Entraining Agent. *Canadian Chemistry and Process Industries, 25,* 13 (abstract from reprint papers numbers 33 and 38).

———. Partial Pressure Processes. *Industrial and Engineering Chemistry, 33,* 1106.

Donald F. Othmer and Charles E. Leyes. Synthesis of Phenol by Partial Pressure Evaporation. *Industrial and Engineering Chemistry, 33,* 158.

Donald F. Othmer and Edward G. Scheibel. Acetone Absorption by Water in a Semi-Commercial Packed Tower. *Transactions of the AIChE, 37,* 211.

Donald F. Othmer and W. Fred Schurig. Destructive Distillation of Maple Wood. *Industrial and Engineering Chemistry, 33,* 188.

Donald F. Othmer and Edward Trueger. Recovery of Acetone and Ethanol by Solvent Extraction. *Transactions of the AIChE, 37,* 597.

Donald F. Othmer and Robert E. White. Condensation of Vapors—Apparatus and Film Coefficients for Lower Alcohols. *Transactions of the AIChE, 37,* 135.

———. New Apparatus Determines Heat Transfer Coefficients for Condensing Vapors of Various Alcohols. *Chemical and Metallurgical Engineering, 48,* 114.

Donald F. Othmer, Robert E. White, and Edward Trueger. Liquid-Liquid Extraction Data. *Industrial and Engineering Chemistry, 33,* 1240.

1942

Raphael Katzen and Donald F. Othmer. Wood Hydrolysis—A Continuous Process. *Industrial and Engineering Chemistry, 34,* 314.

Raphael Katzen, A. O. Reynolds, and Donald F. Othmer. Plastics from Hydrolyzed Lignocellulose. *Transactions of the AIChE, 38,* 735.

Robert R. Lyman and Donald F. Othmer. Scheduling the 168-Hr. Week. *Factory Management and Maintenance, 100,* 73.

———. A Work-Week Plant for Chemical Industry. *Chemical and Metallurgical Engineering, 49,* 86.

Donald F. Othmer. Correlating Vapor Pressure and Latent Heat Data—Use of Critical Constants. *Industrial and Engineering Chemistry, 34,* 1072.

———. Distillation Practices and Methods of Calculation. *Chemical and Metallurgical Engineering, 49,* 84.

———. Oxalic Acid from Sawdust. American Cyanamid and Chemical Corporation (advertisement).

———. Some Modern Chemical Engineering Processes and Their Applications. *Canadian Chemistry and Process Industries, 26,* 15.

Donald F. Othmer, Carl H. Gamer, and Joseph J. Jacobs Jr. Oxalic Acid from Sawdust—Optimum Conditions for Manufacture. *Industrial and Engineering Chemistry, 34,* 262.

Donald F. Othmer, Joseph J. Jacobs Jr., and Joseph F. Levy. Nitration of Benzene—Continuous Process, Using Nitric Acid Alone. *Industrial and Engineering Chemistry, 34,* 286.

Donald F. Othmer, Joseph J. Jacobs Jr., and Arthur C. Pabst. Continuous Fusion Process. *Industrial and Engineering Chemistry, 34,* 268.

Donald F. Othmer and Robert H. Royer. Recovery of Products Resulting from Treatment of Wood with Caustic. *Industrial and Engineering Chemistry, 34,* 274.

Donald F. Othmer and Philip E. Tobias. Liquid-Liquid Extraction Data—Toluene and Acetaldehyde Systems—Tie Line Correlation—Partial Pressure of Ternary Liquid Systems and the Prediction of Tie Lines (3 papers). *Industrial and Engineering Chemistry, 34,* 690, 693, 696.

Donald F. Othmer and Robert E. White. Correlating Gas Solubilities and Partial Pressure Data. *Industrial and Engineering Chemistry, 34,* 952.

Edward G. Scheibel and Donald F. Othmer. A General Method for Calculating Diffusional Operations such as Extraction, Distillation and Gas Absorption. *Transactions of the AIChE, 38,* 339.

———. (Errata) A General Method for Calculating Diffusional Operations such as Extraction, Distillation and Gas Absorption. *Transactions of the AIChE, 38,* 883.

———. Nomographs for Mean Driving Forces in Diffusional Problems. *Industrial and Engineering Chemistry, 34,* 1200.

Robert E. White and Donald F. Othmer. Gas Absorption in a Stedman Packed Column. *Transactions of the AIChE, 38,* 1067.

1943

Robert S. Aries and Donald F. Othmer. Wood in the Chemical Industries. *Timber (Canada),* February.

Roger Gilmont and Donald F. Othmer. Automatic Pressure-Regulating Manometer. *Industrial and Engineering Chemistry, 15,* 641 (Analytical Ed.).

Joseph J. Jacobs and Donald F. Othmer. Deterioration of Lubricating Oils—Soybean Lecithin as an Inhibitor. *Industrial and Engineering Chemistry, 35,* 883.

Joseph J. Jacobs Jr. , Donald F. Othmer, and Allan Hokanson. Sulfonation of Aniline—Application of Partial-Pressure Distillation. *Industrial and Engineering Chemistry, 35,* 321.

Raphael Katzen, Robert E. Muller, and Donald F. Othmer. Destructive Distillation of Lignocellulose. *Industrial and Engineering Chemistry, 35,* 302.

Donald Q. Kern and Donald F. Othmer. Effect of Free Convection on Viscous Heat Transfer in Horizontal Tubes. *Transactions of the AIChE, 39,* 517.

Donald F. Othmer. Alcohol for War. *The Nebraska Blue Print, 42,* 74.

———. Composition of Vapors from Boiling Binary Solutions. *Industrial and Engineering Chemistry, 35,* 614.

Donald F. Othmer and Saul Berman. Condensation of Vapors—Film Coefficients for Alcohols, Esters, and Ketones. *Industrial and Engineering Chemistry, 35,* 1068.

Donald F. Othmer and George A. Fernstrom. Destructive Distillation of Bagasse. *Industrial and Engineering Chemistry, 35,* 312.

Donald F. Othmer, Joseph J. Jacobs Jr., and Wilbur J. Buschmann. Sulfonation of Naphthalene—Application of Partial-Pressure Distillation. *Industrial and Engineering Chemistry, 35,* 326.

Donald F. Othmer and Raphael Katzen. Tar Elimination in Pyroligneous Acid—Removal of Tar-Forming Constituents by Polymerization and Condensation. *Industrial and Engineering Chemistry, 35,* 288.

Donald F. Othmer and Robert L. Ratcliffe Jr. Alcohol and Acetone by Solvent Extraction. *Industrial and Engineering Chemistry, 35,* 798.

Donald F. Othmer and Frederick G. Sawyer. Correlating Absorption Data—Temperature, Pressure, Concentration, Heat. *Industrial and Engineering Chemistry, 35,* 1269.

Alfred F. Schmutzler and Donald F. Othmer. Soybean Protein Dispersions as Printing Ink Vehicles. *Industrial and Engineering Chemistry, 35,* 1196.

Theodore O. Wentworth, Donald F. Othmer, and George M. Pohler. Absolute Alcohol, An Economical Method of Its Manufacture II Plant Data. *Transactions of the AIChE, 39,* 565.

1944

Robert F. Benenati and Donald F. Othmer. Film Coefficients for the Condensation of Vapors. *Chemical and Metallurgical Engineering, 51,* 107.

Roger Gilmont and Donald F. Othmer. Composition of Vapors from Boiling Binary Solutions Water Acetic Acid System at Atmospheric and Subatmospheric Pressures. *Industrial and Engineering Chemistry, 36,* 1061.

Donald F. Othmer. Correlating Vapor Pressure and Equilibrium Constant Data. *Industrial and Engineering Chemistry, 36,* 669.

———. Development of Chemical Equipment. *Canadian Chemistry and Process Industries,* July.

Donald F. Othmer and Robert F. Benenati. Gas Solubility and Partial Pressure—Nomograph for Correlation of Data. *Industrial and Engineering Chemistry, 36,* 375.

Donald F. Othmer and Roger Gilmont. Correlating Vapor Compositions and Related Properties of Solutions. *Industrial and Engineering Chemistry, 36,* 858.

Donald F. Othmer and Hugo L. Kleinhans Jr. Nitration of Toluene—Continuous Partial Pressure Process Using Nitric Acid Alone. *Industrial and Engineering Chemistry, 36,* 447.

Donald F. Othmer, Robert C. Kollman, and Robert E. White. Gas-Liquid Solubilities and Pressures in Presence of Air Acetone-Water and Acetone-Hydrocarbon Oil Systems. *Industrial and Engineering Chemistry, 36,* 963.

George Papps and Donald F. Othmer. Tall Oil—Separation of Stearic and Abietic Acid by Selective Adsorption after Hydrogenation. *Industrial and Engineering Chemistry, 36,* 430.

Frederick G. Sawyer and Donald F. Othmer. Adsorption of Solvent Vapors on Commercial Activated Carbon. *Industrial and Engineering Chemistry, 36,* 894.

Edward G. Scheibel and Donald F. Othmer. Gas Absorption as a Function of Diffusitives and Flow Rates Recovery of Methyl Ketones from Air. *Transactions of the AIChE, 40,* 611.

———. Hydrocarbon Gases—Specific Heats and Power Requirements for Compression. *Industrial and Engineering Chemistry, 36,* 580.

Alfred F. Schmutzler and Donald F. Othmer. Printing Inks from Colloidal Dispersions of Soybean Protein. *Industrial and Engineering Chemistry, 36,* 847.

1945

Raphael Katzen, Robert S. Aries, and Donald F. Othmer. Wood Hydrolysis—Utilization of Waste Liquors. *Industrial and Engineering Chemistry, 37,* 442.

Raphael Katzen, Frederick G. Sawyer, and Donald F. Othmer. Extraction of Lignin from Hydrolyzed Lignocellulose. *Industrial and Engineering Chemistry, 37,* 1218.

Charles E. Leyes and Donald F. Othmer. Continuous Esterification of Butanol and Acetic Acid, Kinetic and Distillation Considerations. *Transactions of the AIChE, 41,* 157.

———. Written discussion on "Continuous Esterification of Butanol and Acetic Acid, Kinetic and Distillation Considerations." *Transactions of the AIChE, 41,* 481.

———. Esterification of Butanol and Acetic Acid. *Industrial and Engineering Chemistry, 37,* 968.

Donald F. Othmer. Using an Orifice to Assist Control of Low Pressure Steam Flow. *Chemical and Metallurgical Engineering, 52,* 122.

Donald F. Othmer and Robert F. Benenati. Composition of Vapors from Boiling Binary Solutions—Aqueous Systems of Acetone, Methanol, and Methyl Ethyl Ketone; and Other Systems with Acetic Acid as One Component. *Industrial and Engineering Chemistry, 37,* 299.

Donald F. Othmer, William S. Bergen, Nathan Shlechter, and Paul F. Bruins. Liquid-Liquid Extraction Data—Systems Used in Butadiene Manufacture from Butylene Glycol. *Industrial and Engineering Chemistry, 37,* 890.

Donald F. Othmer and John W. Conwell. Correlating Viscosity and Vapor Pressure of Liquids. *Industrial and Engineering Chemistry, 37,* 112.

Donald F. Othmer and John G. Harrison Jr. Alinement Chart for Tensile Strength of Rubber and Plastic Compounds. *Rubber World, 113,* 84.

Donald F. Othmer, Nathan Schlechter, and Walter A. Koszalaka. Composition of Vapors from Boiling Binary Solutions—Systems Used in Butadiene Manufacture from Butylene Glycol. *Industrial and Engineering Chemistry, 37,* 895.

Nathan Shlechter, Donald F. Othmer, and Robert Brand. Pyrolysis of 2,3-

Butylene Glycol Diacetate to Butadiene. *Industrial and Engineering Chemistry*, *37*, 905.

Nathan Shlechter, Donald F. Othmer, and Seymour Marshak. Esterification of 2,3-Butylene Glycol with Acetic Acid. *Industrial and Engineering Chemistry*, *37*, 900.

1946

Robert S. Aries, R. Laster, and Donald F. Othmer. Carotene. *The Chemurgic Digest*, *5*, 255.

Donald F. Othmer and Samuel Josefowitz. Correlating Viscosities of Gases with Temperature and Pressure. *Industrial and Engineering Chemistry*, *38*, 111.

Donald F. Othmer and Arthur H. Luley. Correlating Equilibrium Constants—Chemical Reactions and Heats of Reaction. *Industrial and Engineering Chemistry*, *38*, 408.

Donald F. Othmer and F. Robert Morley. Composition of Vapors from Boiling Binary Solutions, Apparatus for Determination under Pressure. *Industrial and Engineering Chemistry*, *38*, 751.

George Reithof, Sidney G. Richards, Sidney A. Savitt, and Donald F. Othmer. Method for Purifying Beta-Picoline and a Test for Purity. *Industrial and Engineering Chemistry*, *18*, 458 (Analytical Ed.).

1947

Henry R. Linden and Donald F. Othmer. Combustion Calculations for Hydrocarbon Fuels—Part I. *Chemical Engineering*, *54*, 115.

———. Combustion Calculations for Hydrocarbon Fuels—Part II. *Chemical Engineering*, *54*, 148.

Donald F. Othmer and Samuel Josefowitz. Composition of Vapors from Boiling Binary Solutions Acetonitrile-Water System. *Industrial and Engineering Chemistry*, *39*, 1175.

1948

Saul Bergman, Hrant Isbenjian, Arthur Sedoff, and Donald F. Othmer. Esterification—Continuous Production of Dibutyl Phthalate in a Distillation Column. *Industrial and Engineering Chemistry*, *40*, 2130.

Saul Bergman, Andrew A. Melnychuck, and Donald F. Othmer. Dibutyl Phthalate—Reaction Rate of Catalytic Esterification. *Industrial and Engineering Chemistry*, *40*, 1312.

Samuel Josefowitz and Donald F. Othmer. Adsorption of Vapors—A New Apparatus, and Data for the Lower Ketones, Benzene, and Hexane on Activated Carbon. *Industrial and Engineering Chemistry*, *40*, 739.

Donald F. Othmer. Composition of Vapors from Boiling Solutions—Improved Equilibrium Still. *Analytical Chemistry*, *20*, 763.

———. Correlaciones entre datos fisicos y quimicos para su utilizacion en ingenieria. *Euclides* (Spanish), *8*, 57.

———. Correlation entre les donnees physiques et chimiques a l'usage de l'industrie chimique. *Chimie et Industrie* (French), *59*, 329.

———. Correlation entre les donnees physiques et chimiques a l'usage de l'industrie chimique. *Chimie et Industrie* (French), *59*, 446.

———. Correlation entre les donnees physiques et chimiques a l'usage de l'industrie chimique. *Chimie et Industrie* (French), *59*, 552.

Donald F. Othmer and Roger Gilmont. Correlating Vapor Compositions and Related Properties of Solutions—Use of Critical Constants. *Industrial and Engineering Chemistry*, *40*, 2118.

Donald F. Othmer and Samuel Josefowitz. Correlating Adsorption Data. *Industrial and Engineering Chemistry*, *40*, 723.

Donald F. Othmer, Samuel Josefowitz, and Alfred E. Schmutzler. Correlating Densities of Liquids—A New Nomograph. *Industrial and Engineering Chemistry*, *40*, 883.

———. Correlating Surface Tensions of Liquids—A New Nomograph. *Industrial and Engineering Chemistry*, *40*, 886.

Donald F. Othmer and Sidney A. Savitt. Composition of Vapors from Boiling Binary Solutions—Systems of Phenol with Beta- and Gamma-Picolines and 2,6-Lutidine. *Industrial and Engineering Chemistry*, *40*, 168.

Donald F. Othmer and Salvatore J. Silvis. Correlating Boiling Point Elevations of Sugar Solutions as a Function of Pressures, Concentrations and Percent Purity: A New Nomograph. *Sugar*, *43*(7), 28.

———. Correlating Viscosity, Temperature and Concentration of Sucrose Solution. *Sugar*, *43*(5), 32.

Alfred Schmutzler and Donald F. Othmer. Chromium-Rhodamine B Toners. *American Ink Maker*, *26*, 28.

1949

Robert S. Aries and Donald F. Othmer. Methods of Determining Plant Location. *Chemical Engineering Progress*, *45*, 285.

Frederick G. Eichel and Donald F. Othmer. Benzaldehyde by Autoxidation of Dibenzyl Ether. *Industrial and Engineering Chemistry*, *41*, 2623.

H. R. Linden and Donald F. Othmer. Air Flow through Orifices in the Viscous Region. *Transactions of the ASME*, *71*, 44 (abstract) preprinted as paper # 48-A-93 (Nov.–Dec. 1948).

Donald F. Othmer and Arthur H. Luley. A New Method of Refining Sugar. *Sugar Journal*, *11-12*, 3.

———. Refining Sugar by Solvent. *Food*, *18*, 81.

———. Solvent Refining of Raw Sugar. *Sugar*, *44*, 29.

Donald F. Othmer and E. S. Roskowski. Data—Interpretation and Correlation. *Encyclopedia of Chemical Technology*, *4*, 846.

Donald F. Othmer, Sidney A. Savitt, Alfred Krasner, Alan M. Goldberg, and David Markowitz. Composition of Vapors from Boiling Binary Solutions Systems with m- or p-Cresol as One Component. *Industrial and Engineering Chemistry*, *41*, 572.

Donald F. Othmer and Jose Serrano Jr. Solubility Data for Ternary Liquid Systems—Systems of Acetic Acid, Higher Boiling Homologous Acids, and Water. *Industrial and Engineering Chemistry*, *41*, 1030.

Donald F. Othmer and Edward H. Ten Eyck Jr. Correlating Azeotropic Data. *Industrial and Engineering Chemistry, 41,* 2897.

Sidney A. Savitt, Alan M. Goldberg, and Donald F. Othmer. Colorimetric Analysis of m- and p-Cresol in Their Mixtures. *Analytical Chemistry, 21,* 516.

Alfred E. Schmutzler and Donald F. Othmer. Oil Adsorption by Freshly Precipitated Pigments. *American Ink Maker, 27,* 29.

1950

Roger Gilmont, Eugene A. Weinman, Franklin Kramer, Eugene Miller, Frank Hashmall, and Donald F. Othmer. Thermodynamic Correlation of Vapor-Liquid Equilibria, Determination of Activity Coefficients from Relative Volatility. *Industrial and Engineering Chemistry, 42,* 120.

Raphael Katzen, Robert S. Aries, Kjell Goranson, and Donald F. Othmer. Utilization of Cellulose from Methanol-Chlorination of Wood. *TAPPI, 33,* 246.

Raphael Katzen, Irving Krushel, and Donald F. Othmer. Utilization of Methoxy Lignin Chloride by Resinification. *TAPPI, 33,* 244.

Raphael Katzen, Joseph Pearlstein, Robert Muller, and Donald F. Othmer. A New Process for the Separation of Lignin and Cellulose. *TAPPI, 33,* 67.

Donald F. Othmer and Ju Chin Chu. Review of Process Heat Transfer. *Chemical Engineering, 57,* 303.

Donald F. Othmer and Sanjeev Ananda Rao. n-Butyl Oleate from n-Butyl Alcohol and Oleic Acid. *Industrial and Engineering Chemistry, 42,* 1912.

Donald F. Othmer and Louis G. Ricciardi. Wallboard from Wood Waste without Resins by a Dry Process. *Northeastern Wood Utilization Council Bulletin, 31.*

Donald F. Othmer and Salvatore J. Silvis. Correlating Viscosities—Caustic Soda Solutions. *Industrial and Engineering Chemistry, 42,* 527.

1951

Andrew E. Karr, Edward G. Scheibel, William M. Bowes, and Donald F. Othmer. Compositions of Vapors from Boiling Solutions—Systems Containing Acetone, Chloroform, and Methyl Isobutyl Ketone. *Industrial and Engineering Chemistry, 43,* 961.

W. Paul Moeller, Sven W. Englund, Tsu Kan Tsui, and Donald F. Othmer. Compositions of Vapors from Boiling Solutions, Equilibria under Pressure of Systems: Ethyl Ether-Ethyl Alcohol and Ethyl Ether-Water Ethyl Alcohol. *Industrial and Engineering Chemistry, 43,* 711.

Donald F. Othmer. The Background of Chemical Engineering. *Bombay Technologist, 1,* 3.

———. Chemical Engineering at the Polytechnic Institute of Brooklyn. *Bombay Technologist, 1,* 11.

———. Flash Vaporization Equilibrium Still (advertisement). *Industrial and Engineering Chemistry, 43.* The Emil Greiner Co.

Donald F. Othmer, Robert S. Aries, and Marvin D. Weiss. Water Pollution by Industry. *Mechanical Engineering, 73,* 706.

Donald F. Othmer and Roger Gilmont. Correlating Physical and Thermodynamic Properties. *Petroleum Refiner, 30-31,* 111, 107.

———. Correlating Physical and Thermodynamic Properties Used in Petroleum Technology. Third World Petroleum Congress, Preprint 13, sec. III (Holland–1951) "Proceedings," Sec. III, p. 1.

Donald F. Othmer, W. Paul Moeller, Sven W. Englund, and Robert G. Christopher. Composition of Vapors from Boiling Binary Solutions, Recirculation-Type Still and Equilibria under Pressure for Ethyl Alcohol-Water Systems. *Industrial and Engineering Chemistry, 43,* 707.

Donald F. Othmer and F. Murata. Wallboard from Wood Waste without Resin by Dry Process. *Wood Industry* (Japan), *6,* 523.

Donald F. Othmer, E. H. Ten Eyck, and Stanley Tolin. Equilibrium Flash Vaporization of Petroleum Crude Oils or Fractions—Method and Apparatus for Determination. *Industrial and Engineering Chemistry, 43,* 1607.

———. Equilibrium Flash Vaporization of Petroleum Crudes or Fractions. Third World Petroleum Congress, Preprint 6, sec. III (Holland–1951) "Proceedings," Sec. III, p. 35.

Donald F. Othmer, M. D. Weiss, and Robert S. Aries. The Utilization of Pollutive Wastes in the Process Industries. ASME Process Industries Conference, Preprint. 17 April.

———. Water Pollution by Industry, A Survey of State Legislation and Regulations. ASME Process In-dustries Conference, Preprint. 17 April.

1952

Robert E. Muller and Donald F. Othmer. Ferric Chloride Treatment of Boneblack. *Sugar, 47*(6), 40.

Donald F. Othmer. Beehive Kiln Operation. *Northeastern Wood Utilization Council, Bull. 37.*

———. History and Present Status of the Industry; Continuous Carbonization, the Mellman Retort. *Northeastern Wood Utilization Council, Bull. 37.*

Donald F. Othmer and Robert D. Beattie. Correlating Heat Quantities of Lower Hydrocarbons, Other Gases. *Petroleum Refiner, 31,* 129.

Donald F. Othmer, Manu M. Chudgar, and Sherman L. Levy. Composition of Vapors from Boiling Binary Solutions—Binary and Ternary Systems of Acetone, Methyl Ethyl Ketone and Water. *Industrial and Engineering Chemistry, 44,* 1872.

Donald F. Othmer, Sal J. Silvis, and Albert Spiel. Composition of Vapors from Boiling Binary Solutions—Pressure Equilibrium Still for Studying Water-Acetic Acid System. *Industrial and Engineering Chemistry, 44,* 1864.

Donald F. Othmer and Mahesh S. Thakar. Correlating Solubility and Distribution Coefficient Data. *Industrial and Engineering Chemistry, 44,* 1654.

Sidney A. Savitt and Donald F. Othmer. Separation of m- and p-Cresols from Their Mixtures. *Industrial and Engineering Chemistry, 44,* 2428.

1953

Robert S. Aries and Donald F. Othmer. El Acetileno y el Etileno como Materias Primas para la Industria Quimica. *Euclides* (Spain), *13,* 1.

Ju Chin Chu, H. C. Ai, and Donald F. Othmer. Phenol Synthesis by Direct Oxidation of Benzene. *Industrial and Engineering Chemistry*, *45*, 1266.

Sven W. Englund, Robert S. Aries, and Donald F. Othmer. Caustic Fusion of Sodium Toluenesulfonate *and* Synthesis of Cresol—Sulfonation of Toluene. *Industrial and Engineering Chemistry*, *45*, 189.

Roger Gilmont, Theodore Roszcowski, and Donald F. Othmer. Correlating Total Pressures of Binary Systems—A New Method Using Reference Systems at Corresponding States. *Petroleum Refiner*, *32*, 167.

Donald F. Othmer. Correlating Physical and Chemical Data for Chemical Engineering Use. Presented at XIth International Congress of Pure and Applied Chemistry, London (1947). Printed from "Proceedings," 1953.

———. Recuperacion de acido acetico a partir de soluciones acuosas—como ejemplo de aquellos obtenidos de la desalination de la madera. *Euclides* (Spain), *13*, 57.

Donald F. Othmer, Louis G. Ricciardi, and Mahesh S. Thakar. Composition of Vapors from Boiling Binary Systems—New Methods of Representing and Predicting Equilibrium Data. *Industrial and Engineering Chemistry*, *45*, 1815.

Donald F. Othmer and Mahesh S. Thakar. Nomograph for Predicting Diffusion Coefficients and Correlating Diffusion Coefficients in Liquids. *Industrial and Engineering Chemistry*, *45*, 589.

Edward H. Ten Eyck and Donald F. Othmer. Correlating Equilibrium Flash Vaporization Data. *Petroleum Refiner*, *32*, 151.

1954

B. S. Edwards, F. Hashmall, R. Gilmont, and Donald F. Othmer. Thermodynamic Correlation of Vapor-Liquid Equilibria Prediction of Multicomponent Systems. *Industrial and Engineering Chemistry*, *46*, 194.

F. Meissner, G. Schweidessen, and Donald F. Othmer. Continuous Production of Hexamethylenetetramine. *Industrial and Engineering Chemistry*, *46*, 724.

F. Meissner, G. Wannschaff, and Donald F. Othmer. Continuous Production of Nitrotoluenes. *Industrial and Engineering Chemistry*, *46*, 718.

1955

C. LeRoy Carpenter and Donald F. Othmer. Entrainment Removal by a Wire-Mesh Separator. *AIChE Journal*, *1*, 549.

Sherman L. Levy and Donald F. Othmer. Synthesis of Pyridines. *Industrial and Engineering Chemistry*, *47*, 789.

Donald F. Othmer and Jagdish C. Agarwal. Extraction of Soybeans, Theory and Mechanism. *Chemical Engineering Progress*, *51*, 372.

Donald F. Othmer and Gerhard Frohlich. Correlating Permeability Constants of Gases through Plastic Membranes. *Industrial and Engineering Chemistry*, *47*, 1034.

Donald F. Othmer and Roger Gilmont. Vapor-Liquid Equilibria and Vapor Pressure. *Encyclopedia of Chemical Technology*, *14*, 611.

1956

J. C. Chu, J. Forgrieve, R. Grosso, S. M. Shah, and Donald F. Othmer. Study of Bubbling Performance in Relation to Distillation and Absorption. *AIChE Journal, 3* (March), 1, 16.

Donald F. Othmer. Protecting Tanks from Corrosion. *Chemical Engineering, 63,* 260.

Donald F. Othmer, Kichiro Kon, and Takeo Igarashi. Acetaldehyde by Chisso Process. *Industrial and Engineering Chemistry, 48,* 1258.

Donald F. Othmer and L. M. Naphtali. Correlating Pressures and Vapor Compositions of Aqueous Hydrochloric Acid. *Journal of Chemical Engineering Data, 1,* 1, 6.

1957

Donald F. Othmer, P. W. Maurer, C. J. Molinary, and R. Kowalski. Correlating Vapor Pressures and Other Physical Properties. *Industrial and Engineering Chemistry, 49,* 125.

1958

Donald F. Othmer. Acetic Acid Recovery Methods. *AIChE, 54,* 7, 48.

———. Chemicals Recovery from Pulping Liquors. *Industrial and Engineering Chemistry, 50* (March), 60a.

Donald F. Othmer, Fritz Meissner, and Ernst Schwiedessen. Control Instrumentation Makes Possible Continuous Production of Materials in Explosives Industry. Proceedings of Conference on Instruments and Control and Process Industry, p. 71 (6–7 Feb. 1957, published 1958).

Donald F. Othmer and Mahesh S. Thakar. Glycol Production—Hydration of Ethylene Oxide. *Industrial and Engineering Chemistry, 50,* 1235.

1959

H. M. Muller and Donald F. Othmer. Hydraulics of the Uniflux Fractionating Tray. *Industrial and Engineering Chemistry, 51,* 625.

Donald F. Othmer. A Chemical Engineer Looks at the Citrus Industry. *Chemurgic Digest, 18* (2–3).

———. Gewinnung von Essigsaure und Ameisensaure aus Abwassern der Zellstoff-Herstellung. *Chimie-Inginieur-Technik, 31,* 673.

———. Methoden zur Qewinnung von Essigsaure. *Dechema Monographien,* Band 33, Frankfurt, Germany.

Donald F. Othmer and Walter A. Jaatinen. Extraction of Soybeans, Mechanism with Various Solvents. *Industrial and Engineering Chemistry, 51,* 543.

Donald F. Othmer, Ronald Kowalski, and Leonard M. Naphtali. Correlating Heats of Solution and Vapor-Liquid Equilibria. *Industrial and Engineering Chemistry, 51,* 89.

Donald F. Othmer and Ping Liang Ku. Solubility Data for Ternary Liquid Systems: Acetic Acid and Formic Acid Distributed between Chloroform and Water. *Journal of Chemical Engineering Data, 4*(4), 42.

Donald F. Othmer and David Zudkevitch. Correlating Latent Heats and Entropies of Vaporization with Temperature. *Industrial and Engineering Chemistry, 51,* 791.

1960

Donald F. Othmer. Fluidization of Solids. *Encyclopedia of Science and Technology, 5,* 345.

———. Les methods de recouperation de l'acide acetique. *Chimie et Industrie (France), 84*(6), 896.

———. La separation de l'eau et de l'acide acetique. *Chimie et Industrie* (France), *84*(6), 899.

Donald F. Othmer, James J. Conti, and Roger Gilmont. Composition of Vapors from Boiling Binary Solutions, Systems Containing Formic Acid, Acetic Acid, Water and Chloroform. *Journal of Chemical Engineering Data, 5*(3), 301.

Donald F. Othmer and Gerhard J. Frohlich. Correlating Vapor Pressures and Heats of Solution for Ammonium-Nitrate-Water-System—An Enthalpy Concentration Diagram. *AIChE Journal, 6*(2), 210.

Donald F. Othmer, Roger Gilmont, and James J. Conti. An Adiabatic Equilibrium Still: For More Accurate Vapor-Liquid Equilibrium Data. *Industrial and Engineering Chemistry, 52,* 625.

1961

Roger Gilmont, David Zudkevitch, and Donald F. Othmer. Correlation and Prediction of Binary Vapor-Liquid Equilibria. *Industrial and Engineering Chemistry, 53*(July), 574.

Donald F. Othmer. Chemical Engineering Practice in the Chemical Industry. *Japanese Chemical Engineering* (from Japanese text of *Transactions of the Japanese Chemical Engineering Society*), November.

———. A Drop Saved (Water Use and Desalination). *Land Reborn, 12*(May), 1.

———. Fresh Water from the Ocean. *British Chemical Engineering, 6,* 11.

———. Japanese Chemicals Boom. *Chemical Engineering,* (6 Feb.), 54–58.

Donald F. Othmer and Robert D. Beattie. Vacuum Compression Distillation Column (Design and Pressure Losses, Fractionating). *Industrial and Engineering Chemistry, 53*(Oct.), 779.

Donald F. Othmer, R. F. Benenati, and G. C. Goulandris. Vapor Reheat-Flash Evaporation without Metallic Surfaces. *Chemical Engineering Progress, 57*(1), 47.

Donald F. Othmer and Arthur Homme. Sulfuric Acid—Optimized Conditions in Contact Manufacture. *Industrial and Engineering Chemistry, 53*(Dec.), 979.

1962

Donald F. Othmer. Chemical Engineering Practice in the Chemical Industry. *Chemical and Process Engineering* (London: Lomond Technical Press), April.

———. Sea Water Conversion—Some Possible Trends in Future Developments. *Dechema-Monographien* (Frankfurt, Germany), Band 47.

———. Susswasser aus dem Meer. *Dechema-Monographien* (Frankfurt, Germany), Band 41, 143.

Donald F. Othmer, R. F. Benenati, and G. C. Goulandris. Sea Water Desalination without Metallic Heat Transfer Surfaces. *Chemical and Process Engineering* (London: Lomond Technical Press), 564.

———. Vapor Reheat-Sea Water Desalination without Metallic Heat Surfaces. *Dechema-Monographien* (Frankfurt, Germany), Band 47.

Donald F. Othmer and Hung Tsung Chen. Correlating Diffusion Coefficients in Binary Gas Systems. *Industrial and Engineering Chemistry, 1*(Oct.), 249.

Donald F. Othmer and Ning Hsing Chen. New Generalized Equation for Gas Diffusion Coefficient. *Journal of Chemical Engineering Data,* 7, 1.

1963

Donald F. Othmer. Azeotropic Separation. *Chemical Engineering Progress, AIChE, 59*(6), 67–78.

———. Azeotropy and Azeotropic Distillation. In *Kirk-Othmer Encyclopedia of Chemical Technology*, vol. 2, second edition, p. 839.

———. Desalting of Sea Water. *Chemical Engineering, 70* (10 June), 205.

———. Manufactured Water for Agriculture? *Chemurgic Digest, 21*(8), 2.

Donald F. Othmer, R. F. Benenati, and G. C. Goulandris. Vapor Reheat Process for Water Desalination. *Chemical Engineering Progress, AIChE, 59*(12), 63–68.

Donald F. Othmer and William Goldberger. The Kinetics of Nickel Carbonyl Formation. *Industrial and Engineering Chemistry, 2*(3), 202.

Donald F. Othmer and Takeshi Utsumi. Chemical Reaction in Relation to Diffusion Phenomena. *Chemical and Process Engineering* (London: Lomond Technical Press), *44*(8), 419.

1964

Donald F. Othmer. Aridity and Man (book review). *Chemurgic Digest, 22*(4), 2.

———. Chemurgy—The Interrelation of Chemical Industry and Agriculture. *Chemurgic Digest, 22*, 3, 5.

———. Economie—production d'eau douce à partir de l'eau de mer. *Chimie et Industrie* (France), *92*(July), 1, 3.

——— (Consulting Engineering Editor, Div. of U. N.). Technology of Water Desalination—Report for Natural Resources. United Nations Publication #64, B.5, 305.

Donald F. Othmer and Hung Tsung Chen. An Equation of State for Gas Mixtures. *Chemical Engineering Progress.*

Donald F. Othmer and Ning-Hsing Chen. Correlations for the Coefficients of Gaseous Diffusion. I&EC Fundamentals, *3*(3), 279.

Donald F. Othmer and Gerhard J. Frohlich. Carbon Dioxide and Ammonia in Aqueous Ammonium Nitrate Solutions. *Industrial and Engineering Chemistry, 3*(3), 270.

Donald F. Othmer and R. F. Schwab. Dust Explosions. *Chemical and Process Engineering* (London), *45*(4), 165.

Donald F. Othmer and Takeshi Utsumi. Chemical Reaction in Relation to Diffusion Phenomena. CHISA-Congress on Chemical Engineering, Equipment Design and Automation, Prague.

1965

Donald F. Othmer. L'acetylene—augmentation des ventes, consequence d'un meilleur procédé de fabrication. *Chimie et Industrie* (France), *93*(3), 205.

———. Data Interpretation and Correlation. *Kirk-Othmer Encyclopedia of Chemical Technology*, vol. 7, second edition, p. 705.

———. Desalination—Water for the World. *Journal of the Sanitary Engineering Division, Proceedings of the American Society of Civil Engineering*, SA1, *92*(Feb.), 15.

———. Desalination—Worldwide Needs and Opportunities. *Chemical and Process Engineering* (London), *46*(3), 120.

———. Make 3 to 5¢ Acetylene? *Hydrocarbon Processing, 44*(3), 145.

———. Olefins vs. Acetylene—Competitive Raw Materials for the Petrochemical Industries in Developing Countries. *Chemical Age of India, 16*(3), 101.

———. The Possible Impact of Low Cost Acetylene. *Chemical Engineering Progress—Trends in Petrochemicals, AIChE*, (March), 16.

———. Resources of the World. *Chemical and Process Engineering* (London), *46*(Sept.), 484.

Donald F. Othmer and Hwa-Nan Huang. Correlating Vapor Pressure and Latent Heat Data. *Industrial and Engineering Chemistry, 57*(10), 42.

Donald F. Othmer and Takeshi Utsumi. Analog Control—Possibilities and Applications. *Dechema-Monographien* (Frankfurt, Germany), *53*(51), 72.

———. Chemical Reactor Analysis—Co-Current Flow and Counter-Current Flow in Tubes. AIChE, 58th Annual Meeting, Philadelphia, Preprint 39A, December.

1966

Donald F. Othmer. Desalination—Present and Future Importance. *Revista del Colegio de Quimicos de Puerto Rico, 25*(1), 1.

———. Evaporation des liqueurs de fabrication de la pâte à papier par le procédé de rechauffage de la vapeur. *Genie Chimique, 96*(6), 1669.

———. Evaporation for Desalination—Scale Prevention and Removal. *Desalination, 1*(July), 194.

———. Materials for Water Desalination. *Power—Technical Briefs*, (April), 126.

———. Vapor Reheat Evaporation of Pulp Liquors. *Chemical Age of India, 17*(8), 613.

———. Water for the World: Desalination. *Journal of the Sanitary Engineering Division, Proceedings of the American Society of Civil Engineering*, SA1, *93*(Feb.), 293.

Donald F. Othmer and Hung Tsung Chen. An Equation of State for Gas Mixtures. *AIChE Journal, 12*(3), 488.

1967

Donald F. Othmer. Half-Stage May Boost Water Plant Output. *Chemical & Engineering News*, (25 Sept.), 64.

———. Vapor Reheat Evaporation of Pulp Liquors. *TAPPI Journal, 50*(3), 101.

1968

Joseph R. Clark and Donald F. Othmer. Sewage Treatment by Coagulation, Sedimentation, and Adsorption. AIChE Symposium Series, *68*, 120, 105 ("Advances in Separation Techniques").

Donald F. Othmer. Condensation Coefficient of Heat Transfer. *Chemical and Process Engineering*, (June).

———. Evaporation for Desalination—Improved Multistage Flash (MSF) Processes. *Desalination*, *5*(Dec.).

———. Expanding the World's Resources (Abridged version of lectures given at the U.S. Army War College). *The Brooklyn Engineer* (three parts), *71*(March, April, May), 6, 7, 8.

———. Expanding the World's Resources (Abridged version of lectures given at the U.S. Army War College). *Science Education*, *52*(March), 2.

———. Expanding the World's Resources. *Chemical Age of India*, *19*(Feb.), 2.

———. Expanding the World's Resources. *Chemurgic Digest*, *25*, *26*, *27* (Jan.–Feb. 1967, 1968, 1969).

———. Heat and Power from Sea Water. *Chemical Age of India*, *19*(June), 6.

———. Heat Transfer for Sewage and Sludge. *Process Biochemistry*, (Feb.), 33–35.

———. Heat Transfer Methods for Sewage and Sewage Sludges. *Dechema-Monographien*, *59*, 1045, 41.

———. Research Opportunities in Desalination. Institute of Hydro-Sciences and Water Technology, pub. no. 20, December.

Donald F. Othmer and Hung-tsung Chen. Correlating and Predicting Thermodynamic Data Reference Substance Equations and Plots. *Industrial and Engineering Chemistry*, "Applied Thermodynamics Symposium," *60*(April), 4.

Donald F. Othmer and Rudolf Nowak. Halide Metallurgy: Novel Separation Techniques from Difficult Ores. *AIChE Symposium Series*, *68*, 120, 141 ("Advances in Separation Techniques").

Donald F. Othmer and Erl-Sheen Yu. Correlating Vapor Pressures and Vapor Volumes Use of Reference Substance Equations. *Industrial and Engineering Chemistry*, *60*(Jan.), 1.

Ruiten Ouyang and Donald F. Othmer. Correlation of Vapor-Liquid Equilibrium Data Using Steric Relation. *Journal of Chemical Engineering of Japan*, *1*(2), 99.

1969

Donald F. Othmer. Desalination of Sea Water. *Encyclopedia of Marine Resources*.

———. Evaporation for Desalination—Improved Multistage Flash (MSF) Processes. *Desalination*, *6*, 13.

———. Heat and Power from Sea Water. *Encyclopedia of Marine Resources*.

———. Our Race for Resources: Can We Win It? *Think* (IBM), *35*(Nov.–Dec.), 6.

———. Pipeline Tracing with SECT May Save Money. *Power* (McGraw-Hill), June.

———. Piping in Electric Heat. *Chemical Week* (McGraw-Hill), *41*(14 June).

Donald F. Othmer and Hung-Tsung Chen. Thermodynamics. In *Kirk-Othmer Encyclopedia of Chemical Technology*, vol. 20, second edition.

Donald F. Othmer, B. M. Fabuss, and Alexander Korosi. Viscosities of Aqueous Solutions of Several Electrolytes Present in Sea Water. *Journal of Chemical and Engineering Data*, *14*(2), 192.

1970

Masao Ando and Donald F. Othmer. Heating Pipelines with Electrical Skin Current. *Chemical Engineering*, 9 March.

Joseph R. Clark and Donald F. Othmer. Industrial and Domestic Sewage, A Total Treatment with Fly Ash and Polyelectrolytes. *Dechema-Monographien, 64.*

Donald F. Othmer. Evolution et tendances—l'eau—besoins, fourniture et production par dessalement. *Chimie et Industrie—Genie Chimique, 103*(July), 12.

———. Man versus Materials. *Transactions of the New York Academy of Sciences*, Series 11, *32*(March), 3.

———. Water and Life. *Chemistry, 43*(Nov.), 10.

———. Water in Chemical Industry—Supply, Recovery, Reuse. *Dechema-Monographien, 64.*

———. Water in the Chemical Industry. *Chemical and Process Engineering* (London), (June), 142.

———. Water—Requirements, Supply and Production by Desalination. In *Kirk-Othmer Encyclopedia of Chemical Technology*, vol. 22, second edition, October.

Ralph C. Roe and Donald F. Othmer. Controlled Flash Evaporation—an Improved Multi-stage Flash System. Third International Symposium on Fresh Water from the Sea, *1*, 169.

1971

Donald F. Othmer. Man—Materials—Midden. *Chemist, 48*(Feb.), 2.

———. Man—Materials—Midden. *The Hexagon of Alpha Chi Sigma*, Spring.

———. Pipe Line Heating. In *Kirk-Othmer Encyclopedia of Chemical Technology*, supplement, vol. 23, July.

Donald F. Othmer and John Griemsmann. Arctic Pipeline Can Be Heated. *Pipeline and Gas Journal, 198*(8), 38.

———. Moving the Arctic Oil: Pipelines and the Pour Point. *Mechanical Engineering*, November.

Donald. F. Othmer and Rudolf Nowak. Chlorination Process Upgrades Low Grade Ores. *Chemical & Engineering News*, 29 November.

Donald F. Othmer and Ralph C. Roe. Controlled Flash Evaporation. *Chemical Engineering Progress, 67*(July).

Ralph C. Roe and Donald F. Othmer. Controlled Flash Evaporation. *Mechanical Engineering*, May.

1972

Leslie L. Balassa and Donald F. Othmer. Use of Azeotropic Distillation. Advances in Chemistry Series, No. 115, "Extractive and Azeotropic Distillation," June.

Joseph R. Clark and Donald F. Othmer. A Simple Total Sewage Treatment by Coagulation, Sedimentation, and Adsorption. *AIChE Symposium Series, 68*, 120.

Donald F. Othmer. Die Bedeutung der synthetischen Elastomere. *Kunststofftechnik, 11*(2), 17.

———. Heating of Steel Pipelines. Eighth Arab Petroleum Congress (Algiers), paper no. 30, 28 May–3 June.

———. Man—Materials—Midden. *The Relevant Scientist, 1*(Nov.), 1.

———. Wasserwirtschaft und chemische Industrie (Water Requirements and Supply for the Chemical Industry). *Wasser, Luft, und Betrieb, 16*(9), 301.

Donald F. Othmer and Hung Tsung Chen. Reference Equation of State and Thermodynamic Properties. Proceedings of the First International Conference on Calorimetry and Thermodynamics (Polish Academy of Sciences, Warsaw).

Donald F. Othmer and Rudolf Nowak. Chlorination Process to Upgrade Low Grade Ores. *Chemical Age of India, 23(2), 155.*

———. Halogen Affinities—a New Ordering of Metals to Accomplish Difficult Separations. *AIChE Journal, 18*, 1.

1973

Donald F. Othmer. Desalination—Methods and Processes. *Iranian Journal of Science and Technology, 2*, 4. (Paper presented at First Iranian Congress of Chemical Engineering, Shiraz, Iran, May 1973).

———. Power, Fresh Water, and Food from Cold, Deep Sea Water. Fourth International Symposium on Fresh Water from the Sea, Heidelberg, 9–14 Sept, *2*, 497.

———. Water for Cities and Industries. *Indian and Eastern Engineer*, 114th Anniversary number.

———. Water Reuse in Industry: Part 5—The Water Pollution Control Act: Reaching toward Zero Discharge. *Mechanical Engineering*, September.

Donald F. Othmer and Rudolf Nowak. TiO2 and Ti-Metal through Chlorination of Ilmenite. *Chemical Engineering Progress, 69*(June), 6.

Donald F. Othmer and Oswald A. Roels. Exploiting Resources of Sea Water. *Indian and Eastern Engineer*, November.

———. Power, Fresh Water, and Food from Cold, Deep Sea Water. *Science, 182*(4108), 121.

1974

Donald F. Othmer. Azeotropic Distillation (part one of number 313a). *Verfahrenstechnik*, no. 3, August.

———. Azeotropic Distillation (Part II: Uses, Examples, History. *Verfahrenstechnik*, no. 4, August.

———. Energy. *Chemical Age of India, 25*(Dec.), 881.

———. Energy. *Mechanical Engineering*, August.

———. Industrial and Domestic Sewage: A Total Treatment with Fly Ash and Polyelectrolytes. In *Wastes—Solids, Liquids, Gases*. New York: Chemical Publishing.

———. Water Pollution Control Act: Reaching toward Zero Discharge. American Industrial Report, No. 7, 4-9, published in Chinese in Mainland China, September.

Donald F. Othmer, Rudolf Nowak, and Lutfi Durak. Chlorine Affinity Series of Metals: Its Uses in Novel Separations from Difficult Ores. Proceedings of the First Iranian Congress of Chemical Engineering.

Donald F. Othmer and Oswald A. Roels. Power, Fresh Water, and Food from Cold, Deep Sea Water. In *Energy, Use, Conservation and Supply.* Washington, D.C.: AAAS.

———. Power, Fresh Water and Food from Cold, Deep Sea Water. *American Industrial Reports, 4*(March).

———. Power, Fresh Water, and Food from Cold, Deep Sea Water. *MTS Journal, 8*(Sept.), 7.

———. Power, Fresh Water, and Food from Cold, Deep Sea Water. *Tracings, 15*(Summer), 3.

1975

Donald F. Othmer. Energy—Alternative Sources to Oil. "An American Viewpoint." *Chemical Age of India, 26*(June), 6.

———. Fresh Water, Energy, and Food from the Sea and the Sun. *Desalination, 17*(Oct.), 193.

1976

Donald F. Othmer. Acetylene of Acetylene—Chemicals from Natural Gas. *American Industrial Report, 5*(May).

———. Digestion of Aqueous Wastes with Oxygen Pressure—Recycling Oxidation Sewage Treatment. *Process Biochemistry*, (June).

———. From the Sea: Food, Power and Fresh Water (about Othmer). *Chemist*, (March).

———. Power, Fresh Water, and Food from the Sea. *Mechanical Engineering*, (Sept.).

1977

Donald F. Othmer. Acetylene and Acetylene—Chemicals from Natural Gas. *Verfahrenstechnik, 11*(June), 6.

———. Energy—Fluid/Fuels from Solid. *Chemical Age of India, 28*(April), 4.

———. Energy—Fluid Fuels from Solids. *Mechanical Engineering, 30*(Nov.).

———. Oxygenation of Aqueous Wastes: The PROST System. *Chemical Engineering*, 20 June.

———. Oxygenation of Aqueous Wastes: The PROST System. *American Industrial Report, 25*(Sept.).

———. Pipe Heating by AC in Steel (Part 1). 3R International, July.

———. Pipe Heating by AC in Steel (Part 2). 3R International, September.

———. Trading Technology. *Chemtech*, October.

———. T'was Easier Then. *The Chemist, 5*(Sept.).

1978

Donald F. Othmer. A Little Learning. *Chemistry and Industry*, 18 March.

———. Peat for Power. *American Industrial Report, 3*(March).

———. Peat for Power. *Mechanical Engineering, 55*(May).

1979

Donald F. Othmer. Earth + Water + Air = Fire: The Wet Air Oxidation (WAO) of Wastes. *Mechanical Engineering*, December.

———. Methanol—A Synthetic Liquid Fuel. *Important for the Future IV*, 1 (published by UNITAR), February.

1980

Donald F. Othmer. Alcohols as Fuels and Chemical Feedstocks. *Chemical Weekly* (Silver Jubilee number).

———. Engineering and Chemical Data Correlation. In *Kirk-Othmer Encyclopedia of Chemical Technology*, vol. 9, third edition, p. 45.

———. Methanol—A Synthetic Liquid Fuel. *American Industrial Report, 1*(Jan.).

———. The Methanol Potential. *Chemical Weekly*, 20 May.

———. Methanol: The Fuel of the Future. Dr. Othmer Unfolds the Story (interview). *Chemical Weekly*, 27 May.

1981

Donald F. Othmer. Alcohols as Fuels and Feedstocks. *The Chemical Engineer*, January.

———. Alternative Energy—What's Really Needed (Othmer interviewed). *Reactor*, 46(April).

———. Desalination. In *Encyclopedia of Environmental Science and Engineering*, second edition, p. 140.

———. Methanol—The Efficient Conversion of Valueless Fuels into a Versatile Fuel and Chemical Feedstock. ASME, 81-PID-1.

———. Pipe Heating by AC Steel. *Journal of Pipelines*, January, 113.

1982

Donald F. Othmer. Distillation—Some Steps in Its Development. In *A Century of Chemical Engineering*, William Furter, ed. New York: Plenum Press, p. 259 (proceedings of symposium held at ACS meeting in Las Vegas, Nevada, 1980).

———. Methanol Is the Best Way to Bring Alaska Gas to Market. *Oil & Gas Journal, 80*(1 Nov.), 44.

———. Methanol—Its Production and Use as a Low Cost Fuel. ASME, 82-PET-28.

Theodore O. Wentworth and Donald F. Othmer. Producing Methanol for Fuels. *CEP: Chemical Engineering Progress*, August.

1983

Donald F. Othmer. Azeotropic and Extractive Distillation. *AIChE Symposium Series*, *79*, 235.

———. Methanol—Some Developments in a Versatile Synfuel. ASME, 83-PET-4.

———. Random Reminiscences of Problems in Solvent Extraction (foreword). In *Handbook of Solvent Extraction*. Lo, Baird, and Hanson, eds. New York: Wiley-Interscience.

Donald F. Othmer and M. S. Thakar. Correlating Diffusion Coefficients in Liquids. Current Contents, *14* (7 Nov.), 45 (citation from *Ind. and Eng. Chem. 45* [1953], 589).

1984

Donald F. Othmer. Alcohol Fuel. In *Encyclopedia of Science and Technology*, sixth edition. New York: McGraw-Hill.

———. The Chemical Engineer's Large Future Task—Energy Conversions and Synthetic Fuels. *The Souvenier* (commemorating the first teaching of chemical engineering in India at Jadavpur University, Calcutta, 60 years ago) (reworking of published paper appearing in *A Century of Chemical Engineering*, W. Furter, ed., 1982).

———. Methanol Economical Production from Low Value Solid Fuels. *Chemical Industry News* (Bombay), September.

1985

Donald F. Othmer. Methanol: Fuel for Automobiles. *CEP*, October.

1986

Donald F. Othmer. Alcohols—Fuels for Automobiles. *Chemical Industry News*, May.

———. Alcohols—Fuels for Automobiles. Methanol: An Alternative Fuel. Proceedings of ASME Region V Conference, Reg.V, *1*, 21.

———. Alcohols—Fuels for Automobiles. Technical Economics, Synfuels and Coal Energy—1986, ASME Symposium on "Synfuels and Coal, Energy," 1–7, PD-*5*.

1987

Donald F. Othmer. A High Octane Enhancer or Gasoline Additive from Coal. Technical Economics, Synfuels and Coal Energy—1987, ASME, PD, *8*, 91.

———. Methanol from Low Grade Coal—a Superior Auto Fuel. *Chemical Weekly*, 27 October.

———. Methanol Production from High-Sulfur Ohio Coal. *Chemical Weekly*, 20 January.

———. Methanol—the Neglected Fuel. *Chemical Weekly*, 15 September.

1989

Donald F. Othmer. Multiple Catalysts Produce a Synthetic Fuel from Coal. *Chemical Engineering Communications*, *83*, 65.

1992

Frederick A. Zenz and Donald F. Othmer. The Saturation Dilute-Phase Concentration of Matter. *Chemical Engineering Communications*, *116*, 89.

Awards

1958	Tyler Award, American Institute of Chemical Engineers Barber-Coleman Award, American Society for Engineering Education
1962	D. Eng. (honorary), University of Nebraska
1970	Honor Scroll, American Institute of Chemists
1975	Award of Merit, Association of Consulting Chemists and Chemical Engineers Golden Jubilee Award, Illinois Institute of Technology
1977	Chemical Pioneers Award, American Institute of Chemists D. Eng. (honorary), Polytechnic University
1978	Murphree Exxon Award, American Chemical Society Professional Achievement Award, Illinois Institute of Technology D. Eng. (honorary), New Jersey Institute of Technology Perkin Medal, Society of Chemical Industry
1979	Honorary Fellowship, British Institution of Chemical Engineers
1981	Hall of Fame, Illinois Institute of Technology
1987	Mayor's Award of Honor for Science and Technology, New York City
1989	Outstanding Alumnus Award, University of Nebraska Citation for Improvement of the Quality of Life, Borough of Brooklyn Award for Significant Contributions to the Polytechnic University
1991	Founders Award, American Institute of Chemical Engineers

Alec Jordan, founding editor of *Chemical Week*, and Don at the Perkin Medal Award dinner. Don received the Perkin Medal, the chemical industry's most distinguished award for innovation, in 1978.

Notes on Contributors

George Bugliarello, chancellor, Polytechnic University
Warren E. Buffett, chairman, Berkshire Hathaway
Amy Beth Crow, research assistant, Chemical Heritage Foundation
Tranda S. Fischelis, niece of Donald Othmer
Gerhard J. Frohlich, student of Donald Othmer and 1999 president of the American Institute of Chemical Engineers
Susan G. Hamson, archivist, Chemical Heritage Foundation
W. Alec Jordan, founding editor, *Chemical Week*
James Kingsbury, chairman of the board of trustees, Long Island College Hospital, 1976–1989
Barbara J. Kohuth, vice president of Institutional Development, Long Island College Hospital, 1982–1997
Mary D. Seina, niece of Mildred Topp Othmer
Leslie A. Shad, goddaughter of Mildred Topp Othmer
Arnold Thackray, president, Chemical Heritage Foundation